'Responding to climate change requires that [...] woven ethical and political issues that rea[...] that are often diverse and conflicting. Con[...] unique insights into how and why we come [...] change can be resolved, and will be essential reading for students grappling with these challenging questions.'

—Harriet Bulkeley, Professor of Geography, Durham University, UK

'With public debate over climate change frozen over "matters of fact", Mike Hulme offers essays that debate both sides of "matters of concern". The paired essays show reason on both sides of each question. They show what a rational global debate about climate change would actually look like. And they show that such a debate may actually be possible. This is a unique and hopeful book, which belongs in the library of all students and scholars of climate change.'

—Michael Dove, Margaret K. Musser Professor of Social Ecology, Yale University, USA

'Neither apocalyptic, nor passive towards the most challenging problem for humanity, this book opens a real international and interdisciplinary deliberation about responses to climate change. Solving climate warming is more about matters of concern, about different and shared responsibilities, than it is about matters of fact and the mechanics of energy transitions. Hulme's book encompasses different aspects of the ethical and political debates in a pluralistic way, and offers a good basis for understanding argument and action in our polarized democracies, especially for the younger generation.'

—Bernard Reber, Research Director (political sciences), CNRS, Sciences Po, Paris, France

'Hulme and the contributors to *Contemporary Climate Change Debates* see climate change for what it really is: a political problem, not a scientific one. The science is as certain as it's ever going to be. What students need is a framework for understanding how their values—and the values of well-intentioned others who disagree with them—attach to climate science to produce policy. And that is precisely what this book provides.'

—Lynda Walsh, Associate Professor of English, University of Nevada, USA

'This collection of climate change debates constitutes a timely contribution edited by Mike Hulme, one of the most renowned scholars in climate research, who brings together the necessary cross-disciplinary perspectives. Following up his seminal book *Why We Disagree About Climate Change*, Hulme here undertakes the innovative initiative of bringing in voices of both established and emerging scholars, a very promising move for informative and constructive dialogues.'

—Kjersti Fløttum, Professor of Linguistics, University of Bergen, Norway

'At a time when climate change denial has become a deliberate distraction by vested interests rather than a good faith intellectual engagement, this book offers a refreshing take on decision-making amidst complexity. *These* are the climate change debates we need to have—not, "is it happening?", but "what are we going to do about it?" Mike

Hulme brings together established and emerging scholarly voices in a format that will engage students of many backgrounds.'
—Lesley Head, Professor of Geography, University of Melbourne, Australia

'These carefully designed exchanges by respected scholars allow students to experience meaningful differences of thought and to form their own judgements. Curated by Mike Hulme, a researcher with world-class expertise in scientific and cultural dimensions of climate change, there are no false debates or fake controversies here. Instead, there are mature arguments over questions that will shape future climate pathways. An invaluable classroom resource.'
—Willis Jenkins, Professor of Religious Studies, University of Virginia, USA

'The debates about climate change in this book go far beyond the usual arguments over whether climate change is happening to explore some of the key questions about how to study and attribute change, impacts and costs; whether markets, renewables, solar climate engineering and non-state actors provide the best solutions; and how to address climate justice and communication. I especially appreciate the depth of social science perspectives and the effort to include many voices of women scholars. A terrific resource for teaching, and for researchers wishing to broaden their understanding of key contemporary topics in climate change.'
—Diana Liverman, Regent's Professor of Geography and Development, University of Arizona, USA

'Studying climate change as a student can be daunting. With such a vast array of different literature available it is easy to lose sight of the bigger picture. This volume does justice to the truly interdisciplinary nature of climate change. Written in a tone appropriate for all students from A-level upwards, and with extensive offerings of extra reading for each chapter, Hulme's book is an absolute necessity for those seeking answers to the big questions of climate change. If only it had been published while I was studying!'
—Daisy Malton, geography graduate student, University of Cambridge, UK

Contemporary Climate Change Debates

Contemporary Climate Change Debates is an innovative new textbook which tackles some of the difficult questions raised by climate change.

For the complex policy challenges surrounding climate mitigation, adaptation and resilience, structured debates become effective learning devices for students. This book is organised around 15 important questions, and is split into four parts:

- What do we need to know?
- What should we do?
- On what grounds should we base our actions?
- Who should be the agents of change?

Each debate is addressed by pairs of one or two leading or emerging academics who present opposing viewpoints. Through this format the book is designed to introduce students of climate change to different arguments prompted by these questions, and also provides a unique opportunity for them to engage in critical thinking and debate amongst themselves. Each chapter concludes with suggestions for further reading and with discussion questions for use in student classes.

Drawing upon the sciences, social sciences and humanities to debate these ethical, cultural, legal, social, economic, technological and political roadblocks, *Contemporary Climate Change Debates* is essential reading for all students of climate change, as well as those studying environmental policy and politics and sustainable development more broadly.

Mike Hulme is Professor of Human Geography at the University of Cambridge, UK, founding director of the Tyndall Centre for Climate Change Research and Editor-in-Chief of the review journal *WIREs Climate Change*. He is the author of eight books on climate change, including *Why We Disagree About Climate Change* and *Can Science Fix Climate Change?*

Contemporary Climate Change Debates

A Student Primer

Edited by Mike Hulme

Routledge
Taylor & Francis Group
LONDON AND NEW YORK

from Routledge

First published 2020
by Routledge
2 Park Square, Milton Park, Abingdon, Oxon OX14 4RN

and by Routledge
52 Vanderbilt Avenue, New York, NY 10017

Routledge is an imprint of the Taylor & Francis Group, an informa business

British Library Cataloguing-in-Publication Data
A catalogue record for this book is available from the British Library

Library of Congress Cataloging-in-Publication Data
A catalog record has been requested for this book

ISBN: 978-1-138-33299-7 (hbk)
ISBN: 978-1-138-33302-4 (pbk)
ISBN: 978-0-429-44625-2 (ebk)

Typeset in Times New Roman
by Swales & Willis, Exeter, Devon, UK

To Gill, who indirectly inspired this book through her enthusiastic and exemplary use of philosophy for children (P4C) and who continues to provide me with great support and encouragement.

Contents

Figures

Tables

Acknowledgements

I would like to thank Hannah Ferguson at Routledge/Taylor & Francis who first invited me to consider editing a book such as this on climate change debates. My initial proposal was then commented on by seven external reviewers, whose suggestions were valuable in refining the proposal and concept. Dr Shinichiro Asayama, my former colleague at Cambridge—now of Waseda University, Japan—provided valuable suggestions early on in the project with regards to my proposed topics and authors. Elizabeth Koebele, Assistant Professor of Political Science at University of Nevada, Reno, read through two of the draft chapters and offered helpful suggestions. Geography staff and students at The Perse School, Cambridge, UK, provided helpful feedback on early drafts of two chapters. In particular, teachers James Riley, Peter Hicks, Katie Banks and Ellie Thorne, and students Holly Jones, Lucy Hopes, Katie Russell, Annis Williams, Matthew Chan, Jonathan Black and Florence Brimblecombe are thanked for their feedback. Bill Johncocks produced with supreme efficiency an index of the highest calibre for the book and also helped me correct some egregious inconsistencies in the manuscript. I also thank Daisy Malton, University of Cambridge geography graduate, who read through all the final drafts and helped me construct the glossary and list of acronyms.

Finally, let me thank the production team of Annabelle Harris, Matthew Shobbrook and Hannah Ferguson at Routledge who ensured that everything ran smoothly from initial conception to final appearance. My biggest thanks of course are to the 36 contributing authors to the book who enthusiastically entered into the spirit of the project and, *pro bono*, provided drafts of their various argumentative positions and who worked cooperatively with me through several versions before alighting on the versions that you read here. Individual acknowledgements from any of these authors are appended to the relevant chapters.

Contributors

Liliana B. Andonova, is Professor of IR/Political Science, Head of Inter-disciplinary Programmes and Co-Director of the Centre for International Environmental Studies at the Graduate Institute of International and Development Studies, Geneva, Switzerland. Her research focuses on international institutions, climate change, public-private partnerships, environmental governance and European integration. Andonova is the author of *Governance Entrepreneurs: International Organizations and the Rise of Global Public-Private Partnerships* (Cambridge, 2017) and co-author of *Transnational Climate Change Governance* (Cambridge, 2014).

James Annan is a Director of Blue Skies Research Ltd, an independent consultancy in the UK undertaking academic research in climate change. Prior to this he worked at the Proudman Oceanographic Laboratory in the UK and the Frontier Research Institute for Global Change in Japan, and was a Contributing Author to the Fourth Assessment Report of the IPCC.

Rose Cairns is a Research Fellow at the Science Policy Research Unit, and member of the ESRC STEPS Centre, at the University of Sussex, UK. Between 2012 and 2014 she was a member of the Climate Geoengineering Governance Research Project. Cairns' research focuses on issues of environmental governance and sustainability across a range of domains, with a particular focus on the politics of environmental discourse.

Kim Coetzee is a postdoctoral research fellow in the University of Cape Town's Environmental and Geographical Sciences Department, South Africa. She is currently conducting research on middle class attitudes towards sustainable consumption in relation to food in developing countries. Coetzee has been a delegate at four COPs and received her PhD in 2016 for research on India's role in the UNFCCC negotiations. Her research is at the nexus of consumption, energy transitions and international environmental governance.

John Cook is a Research Assistant Professor at the Center for Climate Change Communication at George Mason University, USA. His research focus is understanding and countering misinformation about climate change. He founded the website Skeptical Science, which won the 2011 Australian Museum Eureka Prize for the Advancement of Climate Change Knowledge. Cook co-authored the books *Climate Change: Examining the Facts*, *Climate Change Science: A Modern Synthesis* and *Climate Change Denial: Heads in the Sand*. More information can be found at http://sks.to/johncook.

Sarah E. Cornell is Associate Professor in sustainability science at Stockholm University, Sweden, researching global change issues at the Stockholm Resilience Centre. Her work focuses on methods to integrate social and biophysical research in an Earth system framing, and on transdisciplinary initiatives at the science-policy-business interface. She is the lead editor of the book *Understanding the Earth System* (Cambridge, 2012).

Michel Crucifix was trained as a physicist and gained expertise in climate and palaeo-climate dyanamics, with a focus on climate models, dynamical systems theory and Bayesian inference. He is a Research Director at the Belgian National Fund of Scientific Research, and Professor at Université Catholique de Louvain, Belgium, where he leads the School of Physics.

Christiane Fröhlich is a Research Fellow at GIGA German Institute of Global and Area Studies in Hamburg, Germany. Her work on the relation between human mobility, global environmental change and violent conflict has been published among others in the journals *Political Geography*, *Mobilities*, *Global Policy*, *Sustainability* and several edited volumes. More information can be found at www.christianefroehlich.de.

Reyer Gerlagh is Professor of Economics at Tilburg University, The Netherlands. He studies the effect of climate change policy on innovation and economic growth, using theory models, empirical models and simulation models. Gerlagh is the author of many articles published in peer-reviewed international journals, ranging from general-interest economics journals to natural sciences journals. He was a coordinating lead author for the IPCC's Fifth Assessment Report (WGIII, Chapter 5), and has held various journal associate editorships. More information can be found at www.gerlagh.nl.

Aarti Gupta is Associate Professor of Global Environmental Governance at Wageningen University, the Netherlands. She is also Lead Faculty and a member of the Scientific Steering Committee of the international Earth System Governance research alliance. Her research focuses on global environmental and climate politics, as well as anticipatory governance of novel technologies. Publications include, *inter alia*, the co-edited volume *Transparency in Global Environmental Governance: Critical Perspectives* (MIT Press, 2014).

Paul G. Harris (www.paulgharris.net) is a political scientist and the Chair Professor of Global and Environmental Studies at the Education University of Hong Kong. He is author or editor of more than 20 books on global environmental and climate change politics, policy and ethics, including *Global Ethics and Climate Change* (Edinburgh University Press, 2016), the *Routledge Handbook of Global Environmental Politics* (Routledge, 2016) and *What's Wrong with Climate Politics and How to Fix It* (Polity, 2013).

Roweno Heijmans is (at the time of writing) a PhD student at Tilburg University, the Netherlands, where he splits his time between working on environmental economics and the biological evolution of preferences. Heijmans' work on the European Union Emission Trading System has appeared in *Nature Climate Change*. Reyer Gerlagh is one of his PhD supervisors—they disagree on nearly everything, but get along very well.

Mike Hulme is Professor of Human Geography at the University of Cambridge, UK, and director of studies for geography at Pembroke College. He is the author of eight books on climate change, including *Why We Disagree About Climate Change* (Cambridge, 2009) and *Can Science Fix Climate Change?* (Polity, 2014). From 2000 to 2007 Hulme was the Founding Director of the Tyndall Centre for Climate Change Research and since 2008 has been the founding Editor-in-Chief of the review journal *WIREs Climate Change*.

Tobias Ide is DECRA Research Fellow at the School of Geography, University of Melbourne, Australia, and an adjunct associate professor of International Relations, Brunswick University of Technology. Ide has published widely on climate change, environmental stress, peace and conflict, including in the journals *Nature Climate Change*, *Global Environmental Change*, *Journal of Peace Research*, *Global Environmental Politics* and *Political Geography*.

Timothy Laing is a Senior Lecturer at the Brighton Business School at the University of Brighton, UK. His research focuses on understanding the challenges of environmental protection and economic development, including climate change policy, low-carbon development, tropical deforestation and the extractive industries. Laing has published in a wide-range of journals including *Ecological Economics*, *Environmental Conservation*, *Journal of Development Studies* and the *Journal of Rural Studies*.

Lei Liu is Associate Professor at the School of Public Administration, Sichuan University, China. His research mainly focuses on environmental policy and governance issues in China. He holds a PhD degree from Peking University (2011) and has been a Jean-Monnet Fellow at the European University Institute (2013–2014) and a visiting scholar at the Ostrom Workshop of Indiana University (2009–2010). Liu's research has been published in academic journals such as *WIREs Climate Change*, *Climate Policy* and *Cities*.

Jane C.S. Long, now retired, was Associate Director for Energy and Environment at Lawrence Livermore National Laboratory, USA, and Dean of Mackay School of Mines at the University of Nevada, Reno. Dr Long serves as a senior contributing scientist for the Environmental Defense Fund and is a fellow of the American Association for the Advancement of Science, an Associate of the National Academies of Science and a Senior Fellow of the California Council on Science and Technology. Dr Long chaired the Bipartisan Policy Center task force on geoengineering.

Greg Lusk is Assistant Professor of History, Philosophy and Sociology of Science at Michigan State University in the USA. His research engages topics relating to climate change, data and measurement and computer simulation. Lusk is currently Co-PI on an NSF-funded project on the use of big data in atmospheric science. He earned his PhD in History and Philosophy of Science from the University of Toronto, after which he completed a postdoc at the University of Chicago on the interdisciplinary 'Climate Change: Limits of the Numerical' project. More about Greg's research can be found at www.greglusk.com.

Kozo Torasan Mayumi is Professor at Tokushima University, Japan. He graduated from the Graduate School of Engineering at the Department of Applied Mathematics and Physics of Kyoto University. Between 1984 and 1988 he studied bioeconomics at Vanderbilt University under Professor Nicholas Georgescu-Roegen's supervision.

Mayumi works in the field of energy analysis, ecological economics and complex hierarchy theory and is an editorial board member of the journals *Ecological Economics*, *Journal of Economic Structures* and *Ecosystem Services*.

Catriona McKinnon is Professor of Political Theory at the University of Exeter, UK. She is the author of *Climate Change and Future Justice* (Routledge, 2011), and the editor of *The Ethics of Climate Governance* (Rowman & Littlefield, 2015) and *Climate Change and Liberal Priorities* (Routledge, 2011). McKinnon is presently finishing a book on climate change and international criminal law and an introductory book on climate ethics. She was the director of the Leverhulme Programme in Climate Justice at the University of Reading from 2015 to 2019.

Gregory Nemet is Professor of Public Affairs and Environmental Studies at the University of Wisconsin–Madison, USA, in the La Follette School of Public Affairs. He teaches courses in energy systems analysis, policy analysis and international environmental policy. Nemet's research focuses on understanding the process of technological change and the ways in which public policy can affect it. As an Andrew Carnegie Fellow he published *How Solar Energy Became Cheap: A Model for Low-Carbon Innovation* (Routledge, 2019).

Peter North is Professor of Alternative Economies in the School of Environmental Sciences at the University of Liverpool, UK. His research focuses on social and solidarity economies as tools for constructing and rethinking alternative geographies for the Anthropocene. North also studies the politics of climate change and ecologically focused social movements engaged in struggles about the implications of anthropogenic climate change and resource constraints for both humans and the wider ecosystems upon which we depend.

Friederike E.L. Otto is a physicist by training with a PhD in philosophy and is currently acting director of the Environmental Change Institute at the University of Oxford, UK. The journal *Nature* called her 'a veteran' on event attribution and she is the author of over 70 peer-reviewed publications and a popular science book *Angry Weather* (Ullstein, 2019: in German) on the topic. Otto is a lead author on the IPCC sixth assessment report and editor on the annual *BAMS* special journal issue on extreme events.

Warren Pearce is Senior Lecturer in the Institute for the Study of the Human (iHuman), University of Sheffield, UK. His research lies at the intersection of climate science, policy and publics and he is the author of journal articles on climate communication, climate change on social media, local climate policy, scientific consensus and cultural representations of climate science. Pearce has written on these topics for *The Guardian* and *The Conversation*, and is a Contributing Author for the IPCC.

Marjan Peeters is Professor of Environmental Policy and Law at Maastricht University, the Netherlands. Her research covers legal aspects of climate change, procedural environmental rights and the use of economic instruments in environmental law. Peeters has co-edited more than six books related to EU environmental and climate law, among which is *Climate Change Law* (Edward Elgar, 2016; co-editor Daniel A. Farber). She is currently co-editing the *Research Handbook on EU Environmental Law* (Edward Elgar, 2020; co-editor Mariolina Eliantonio).

Qing Pei is an Assistant Professor in the Department of Social Sciences at the Education University of Hong Kong, Hong Kong. His research interests include environmental and historical geography, environmental humanities and geo-statistics, work which has been published in journals such as *Annals of the American Association of Geographers, Environmental Research Letters* and *Environmental History*. More information can be found on https://oraas0.ied.edu.hk/rich/web/people_details.jsp?pid=123888.

Marie-Catherine Petersmann is a Postdoctoral Research Fellow, funded by the Swiss National Science Foundation, at the Copernicus Institute of Sustainable Development at Utrecht University, the Netherlands, where she works as part of the Earth System Governance project. Her research focuses on regime interactions and adjudication of conflicts between international environmental law and human rights law. She holds a PhD in International Law from the European University Institute in Florence, Italy.

Tianbao Qin is a Luojia Professor and Director of the Research Institute of Environmental Law (RIEL) of Wuhan University in China, where he is also vice-dean of the Law School. Qin is Co-Editor-in-Chief of *Chinese Journal of Environmental Law* (CJEL) and Secretary-General of the Chinese Society of Environmental Law.

Misato Sato is Assistant Professorial Research Fellow at the Grantham Research Institute on Climate Change and the Environment, and Deputy Director of the ESRC Centre for Climate Change Economics and Policy, both at the London School of Economics and Political Science (LSE), UK. Sato's research interests include the design and impacts of climate and energy policies, including carbon markets design, competitiveness impacts and the low-carbon transition of energy intensive industries.

Mike S. Schäfer is Professor of Science Communication at the University of Zürich, Switzerland, and Director of the University's Center for Higher Education and Science Studies (CHESS). His work focuses on mediated and online communication about climate change, biotechnology and artificial intelligence and about science in general. On these issues Schäfer has published widely and is the co-editor of several encyclopedias, including the three-volume *Oxford Encyclopedia of Climate Change Communication*.

Eloise Scotford is Professor of Environmental Law in the Faculty of Laws and Centre for Law and Environment, University College London, UK. Professor Scotford has research expertise across diverse areas of environmental law and co-authors *Environmental Law: Text, Cases and Materials* (2nd edition, OUP 2019). On climate law, she researches adjudicative processes relating to climate change—e.g. Fisher, Scotford & Barritt, 'The Legally Disruptive Nature of Climate Change' (2017)—and climate-related legislation, e.g. Scotford & Minas, 'Probing the Hidden Depths of Climate Law: Analysing National Climate Change Legislation' (2019).

Kenneth Shockley is Professor and Holmes Rolston III Chair of Environmental Ethics and Philosophy at Colorado State University, USA. He is a frequent observer at the annual UNFCCC Conference of the Parties meetings and has published widely in environmental ethics, the ethical dimensions of climate change and ethical theory.

Jennie C. Stephens is Director of Northeastern University's School of Public Policy and Urban Affairs, USA, and Dean's Professor of Sustainability Science & Policy. Her research, teaching and community engagement focus on social-political aspects of renewable energy transformation, gender dynamics in energy and climate justice and energy democracy. Stephens' book *Smart Grid (R)Evolution: Electric Power Struggles* (Cambridge University Press, 2015) explores social and cultural debates about energy system change.

Ellen Vos is Professor of European Union Law at Maastricht University, The Netherlands, co-director of the Maastricht Centre of European Law and head of the Department of International and European law. Vos's main areas of interest are EU law and governance (comitology and agencies), market integration and EU risk regulation (precautionary principle, food safety, pharmaceuticals nanotechnology). She has published extensively in these areas. She supervises (and has supervised) numerous Master's and PhD theses in these areas.

Pu Wang is Associate Professor at the Institutes of Sciences and Development, Chinese Academy of Sciences. His research fields include climate and energy policies and sustainable development strategies. Wang received his PhD degree from Cornell University in 2014 in the field of natural resources. From 2015 to 2017 he was a postdoctoral fellow at the Belfer Center of the Kennedy School of Government, Harvard University, where he conducted research on China's cap-and-trade systems and China-USA cooperation in climate change policies.

David D. Zhang is a Professor in the School of Geographical Sciences, Guangzhou University, Guangzhou, China. His research interests include physical geography and the 'climate-human' relationship in history, work which has been published in journals such as *Science*, *Proceedings of the National Academy of Sciences*, *Global Ecology* and *Biogeography*.

Meng Zhang is currently a PhD Research Fellow at the Centre for Environmental and Energy Law of Ghent University, Belgium, as well as a PhD Researcher at Wuhan University, China. Zhang is working for his PhD on the topic of 'Carbon Capture and Storage in China and the EU: A Comparative Legal Study in the Context of the Paris Agreement Era', dedicating himself to research on climate change and energy security law and policy, especially the legal issues of carbon capture and storage.

Abbreviations

ANT	actor-network theory
APP	ability-to-pay principle
BPP	beneficiaries-pay principle
BSE	bovine spongiform encephalopathy
CAC	command-and-control
CBA	cost-benefit analysis
CCS	carbon capture and storage
CDM	clean development mechanism
CDP	carbon disclosure project
CDR	carbon dioxide removal
CER	certified emissions reduction (certificate)
CJEU	Court of Justice of the European Union
CVM	contingent valuation method
EDMC	Energy Data and Modelling Center (Tokyo)
EITE	energy intensive-trade exposed industries
ETS	emissions trading system
EU	European Union
FDI	foreign direct investment
GBM	gateway belief model
GDP	gross domestic product
GHG	greenhouse gas
GPTs	general-purpose technologies
GSEP	global sustainable electricity partnership
HOME	Hands Off Mother Earth
HRL	human rights law
ICJ	International Court of Justice
ICLEI	local governments for sustainability (network)
ICTs	information and communication technologies
IEL	international environmental law
INUS	*I*nsufficient but *n*on-Redundant, *U*nnecessary but *S*ufficient
IPBES	Intergovernmental Panel on Biodiversity and Ecosystem Services
IPCC	Intergovernmental Panel on Climate Change
JI	Joint Implementation projects
LDCs	less developed countries
MEE	Ministry of Ecology and Environment (China)
MEP	Ministry of Environmental Protection (China)

NAS	National Academy of Sciences (USA)
NDC	nationally determined contribution
NETs	negative emissions technologies
NGOs	non-governmental organisations
NICE	National Institute for Health and Care Excellence (UK)
NPP	net primary production
NRC	National Research Council (of the NAS) (USA)
OHCHR	Office of the High Commissioner for Human Rights (UN)
ppm	parts per million (usually to describe greenhouse gas concentration)
PPP	polluter-pays principle
PR	public relations
PV	photovoltaic
QALY	quality-adjusted life-year
R&D	research and development
RCP	representative concentration pathway (see glossary)
RE	renewable energy
REDD+	reducing emissions from deforestation and forest degradation (plus sustainability benefits)
SCC	social cost of carbon
SIDS	small island developing states
SRM	solar radiation management
UNDP	United Nations Development Programme
UNFCCC	UN Framework Convention on Climate Change

Glossary

Actor-Network Theory (ANT) a theoretical and methodological approach to social theory where everything in the social and material worlds exists in constantly shifting networks of relationships. All the factors involved in a social situation are on the same level, thus, objects, ideas, processes and any other relevant factors are seen as just as important as humans in creating social situations.

Anthropocene a proposed current (geological) epoch in which humans and their societies have become a global geophysical force.

Basic human rights rights deemed inherent to all human beings, regardless of race, sex, nationality, ethnicity, language, religion, or any other status, for example the right to life and liberty, freedom from slavery and torture, freedom of opinion and expression, and the right to work and education.

Carbon Capture and Storage (CCS) describes the process of capturing waste carbon dioxide usually from large point sources, such as a cement factory or biomass power plant, transporting it to a storage site and depositing it where it will not enter the atmosphere, normally an underground geological formation.

The Chinese model (also known as the Beijing Consensus) refers to the political and economic policies of the People's Republic of China that began to be instituted by Deng Xiaoping after Mao Zedong's death in 1976.

Clean fossil fuel technologies designed to enhance both the efficiency and the environmental acceptability of coal, oil and gas extraction, preparation and use.

Climate sensitivity the change in globally averaged surface temperature (GMT) in response to changes in radiative forcing due, for example, to increased levels of greenhouse gases (GHGs). The *equilibrium* climate sensitivity is often calculated as the settled response of the climate system to a doubling of GHG concentrations in the atmosphere.

Command-and-control regulation the direct regulation of an industry or an activity by legislation that states what is permitted and what is illegal. The 'command' is the designation of quality standards by an authority that must be complied with; the 'control' signifies the negative sanctions that may result from non-compliance.

Cosmopolitanism the ideology that all human beings belong to a single community, based on a shared morality.

Covenant of Mayors this network brings together thousands of local governments voluntarily committed to implementing EU climate and energy objectives. The Covenant was launched in 2008 in Europe.

Echo chambers a communicative situation where certain ideas, beliefs or data are reinforced through repetition within a closed system that does not allow for the free movement of alternative or competing ideas or concepts.

Ecological civilisation   a term popular especially in Chinese political culture to capture the scale of change required in response to global climate disruption and social injustice and that is so extensive as to represent another form of human civilisation, one based on ecological principles.

Eco(logical) collapse   a situation where an ecosystem suffers a relatively sudden, drastic and possibly permanent reduction in carrying capacity for all organisms, often resulting in mass extinction.

Environmental determinism   (also known as climatic determinism or geographical determinism) is the study of how the physical environment predisposes individuals, societies and states towards particular development trajectories.

Filter bubble   a state of intellectual isolation in which online users become separated from information that disagrees with their viewpoints; it can result from personalised searches when a website algorithm selectively guesses what information a user would like to see based on information about the user, such as location, past click-behaviour and search history.

Gaia (hypothesis)   the mythological name given to the Earth by James Lovelock in 1979, which emerges from his argument that all living organisms and inorganic material are part of a dynamical system that shapes the Earth's biosphere and that maintains the Earth as a fit environment for life.

General Purpose Technologies (GPTs)   technologies that can affect an entire economy and that have the potential to dramatically alter societies through their impact on pre-existing economic and social structures, for example the steam engine, railways, electricity, the automobile, the Internet and Artificial Intelligence.

Habermasian ideal   the German sociologist Jürgen Habermas's 'ideal speech situation' occurs in a public sphere where individuals, following a set of rules, are able to freely share their views with one another in a process which closely resembles the true participatory democracy.

Human Rights Law (HRL)   the body of law designed to promote human rights on social, regional and domestic levels. International HRL is primarily made up of treaties, agreements between sovereign states intended to have binding legal effect between the parties that have agreed to them and customary international law.

Information deficit model   (sometimes 'the deficit model' of science communication) it attributes public scepticism or hostility to science and technology to a lack of understanding, resulting from a lack of information, and implies that communication should focus on improving the transfer of information from experts to non-experts.

INUS causality paradigm (see INUS acronym)   'A' is claimed to be an INUS cause of 'B' if 'A' and another condition 'C' together result in 'B', although sets of distinct conditions 'D1', 'D2', etc. resulting in 'B' cannot be ruled out. For example, the cigarette thrown carelessly (A) is an INUS cause of the forest fire (B), since together with a strong wind (C), it results in B. But we recognise that the forest fire could also be caused by several other conditions (e.g. rubbing of trees together, a deliberate act of arson, etc).

Kondratieff Wave   is a long-term economic cycle believed to result from technological innovation and produce a long period of prosperity. This theory was founded by Nikolai Kondratieff, a Communist Russia era economist who noticed agricultural commodity and copper prices experienced long-term cycles.

Kyoto Protocol   an international treaty, negotiated under the UN's Framework Convention on Climate Change, that commits state parties to reduce greenhouse gas emissions by stated amounts. It was adopted in Kyoto, Japan, on 11 December 1997 and entered into force on 16 February 2005. There are currently 192 parties to the Protocol.

Nationally Determined Contributions (NDCs) represent the stated voluntary efforts by each country to reduce national emissions of greenhouse gases and adapt to the impacts of climate change. The Paris Agreement requires each Party to prepare, communicate and maintain successive NDCs that it intends to achieve.

Negative (CO_2) emissions result from the use of carbon capture and storage (CCS) technologies, and other carbon dioxide techniques, that remove carbon dioxide from the atmosphere and that more than offset emissions of the gas from fossil fuel combustion.

Net zero (carbon) emissions refers to achieving net zero carbon dioxide emissions by balancing carbon emissions (from various sources) with carbon removal (often through carbon offsetting) or simply eliminating carbon emissions altogether.

New materialism a term coined in the 1990s to describe a theoretical turn away from the persistent dualisms in modern and humanist traditions whose influences are present in much of cultural theory. The new materialist turn might indeed be considered a 'return' to matter in the context of historical materialism's earlier concern for material circumstances and embodied subject formation.

Non-state actors are institutions, groups, corporations or individuals that hold political influence and which are wholly or partly independent of state governments, for example media organisations, business magnates, lobby groups, religious groups and aid agencies.

Overton Window named after Joseph Overton in the 1990s, the Overton Window is a political theory that refers to the range (or 'window') of policies that the public will likely accept. The idea is that any policy falling outside the Window is out of step with public opinion and the current political climate. The term is often used in relation to the political centre ground and whether a certain party has managed to shift it to the left or right.

Oxford Principles a proposed set of guiding principles for the governance of climate engineering from early research through to the point where they may be available for eventual deployment.

Paris Agreement on Climate Change (PACC) the 2015 international climate treaty, drafted in Paris, through which virtually all sovereign nations have agreed voluntary climate action plans, including emissions reduction targets (see NDCs).

Pigovian tax a tax on any market activity that generates negative externalities: i.e., where the costs of the activity are not included in the market price, as in environmental pollution, tobacco consumption or sugary drinks. The tax is intended to correct an undesirable or inefficient market outcome and does so by being set equal to the social cost of the negative externalities.

Precautionary Principle enables decision-makers to adopt precautionary measures when scientific evidence about an environmental or human health hazard is uncertain and the stakes are high. It first emerged during the 1970s and has since been enshrined in a number of international treaties on the environment, especially within the EU.

Regions 20 (R20) Regions of Climate Action is a not-for-profit international organisation, which brings together a number of leading regions and NGOs, the UN, Development Banks, Clean-Tech companies and academics with a view to accelerate subnational infrastructure investments in the green economy.

Representaive Concentration Pathway (RCP) scenarios of the future (usually to the year 2100) that include time series of emissions and concentrations of the full suite of GHGs and aerosols, as well as changes in land use/land cover. The word 'representative' signifies that each RCP provides only one of many possible scenarios that would lead to the specific radiative forcing characteristics. The number following RCP refers to the eventual radiative forcing measured in Watts per square metre—e.g. 2.6 Wm^{-2}, 8.5 Wm^{-2}.

Social movement theory a loosely related set of ideas within the social sciences that seek to explain why new social movements arise and are sustained, the different forms they take, as well as their potential social, cultural and political consequences.

Solar climate engineering the deliberate manipulation of the atmosphere to alter the amount of incoming solar radiation and thus modify climate.

Tragedy of the commons a term used to describe a situation in a shared-resource system—for example, grasslands, oceans, fisheries, the atmosphere—where individual users acting independently according to their own self-interest behave contrary to the common good of all users by depleting or spoiling that resource through their collective action.

Wicked problems social or political problems that are difficult or impossible to solve because of incomplete, contradictory and changing requirements that are often difficult to recognise or define.

Introduction

Why and how to debate climate change

Mike Hulme

Climate change is an idea that attracts considerable disagreement. Yet it is less an argument about whether or not the climate is changing—there is abundant evidence for this fact from many parts of the world and from both scientific and non-scientific sources. Nor is it even centrally an argument about whether or not the activities of modern human societies are causing climates to change or, at the least, contributing significantly to these changes. True, there are different views expressed in answer to *this* question, but the vast weight of scientific evidence points to a clear answer. I suggest climate change attracts argument and dispute because it is an unsettling idea which challenges many of the things people have traditionally assumed about their weather, namely that it is natural. And most of all, climate change is a source of disagreement because of the far-reaching implications of policy interventions that are often advocated in response to the phenomenon. These interventions raise profound questions about how people should exercise responsibility toward others, about the efficacy of individual versus collective action, about the limits of personal freedom and about how people imagine the future.

In one of my earlier books—*Why We Disagree About Climate Change* (Hulme, 2009)—I explained why there are legitimate reasons for well-meaning people to understand, interpret and respond to the realities of climate change in quite different ways. These disagreements, I explained, are chronic since they emerge from deeply held, yet strongly contrasting, human beliefs, ideologies, values and identities. A decade later, the social, ethical and political fissures exposed by climate change would seem to be wider and deeper, not least because the stakes of different courses of action appear to be higher than ever and because over the last decade we have become *more* socially and politically polarised rather than less (Mason, 2018). Climate change continues to provoke new questions that engage people's hopes and fears, their moral reasoning and their notions of ethical behaviour. Rather than simply being a technical problem to be solved—like how to prevent mercury pollution of rivers or how to replace asbestos in buildings—at heart climate change is entangled with eternal questions about human meaning, purpose, responsibility and ethics. In effect, and to use metaphorical language that has become widespread, climate change is a '**wicked problem**' beyond solution (Rittel & Webber, 1973) rather than a 'tame' one.

The questions prompted by climate change are therefore more about matters of concern than they are about matters of fact (see Callison, 2014). Answering them draws much more upon how people imagine themselves *in* the world than about what they know factually *about* the world. It is people's identities and values, their attitudes to technology, their attachment to the idea of 'progress' and about how they understand their relationships with human and non-human others—whether these others be proximate or distant in either time or space. When disputes erupt over climate science—for example whether climate models are accurate; whether climate measurements are reliable; whether climate scientists are trustworthy—the ensuing arguments are very often acting as proxies for other types of disagreements. These might be about attitudes to nature, about the distribution of wealth, about rights and freedoms, about desirable forms of governance.

These value-laden disagreements about climate change are unlikely to be resolved through more scientific knowledge. Even less, through more assertive scientific predictions about the climatic future. This is why the implicit message from advocates of 'the 97% consensus'—that 'the science is settled' and the sooner everyone accepts this the easier it will be to resolve climate change—can be misleading (although see the debate in **Chapter 9**). Many scientific questions about climate change remain—not least about the future course of climate evolution (see **Chapter 2**). But it is also the case that this line of reasoning—a science-first approach to climate change—mistakenly identifies value disputes as knowledge disputes. And failing to recognise that to resolve value disputes requires one to engage in dialogue and debate rather than in simple name-calling leads some public commentators on climate change down a dead-end alley. They seek to discredit and dismiss people who advance credible arguments about climate policies which happen to contradict their own preferences.

One egregious example of this from a few years ago was the calling out of respected advocates for nuclear power—eminent climate scientists such as Jim Hansen and Tom Wigley—as 'climate denialists'.[1] And more recently I have been labelled a 'climate contrarian' by the American climate scientist Michael Mann.[2] Using simple labels to denigrate one's opponent without considering in detail the reasons for their views is a tactic used to 'win an argument' without in fact winning the argument. In contrast, and as explained by Sharman and Howarth in their review of labelling in climate change discourse, 'Focusing on the motivations behind different opinions about climate change is important … [for] … allowing for the identification of common ground between previously polarized individuals, thus creating a thread by which dialog may begin' (Howarth & Sharman, 2015: 249). Calling out your opponent as a climate 'denier' or 'contrarian'—or indeed as a climate 'alarmist' or 'zealot'—does nothing to encourage constructive dialogue (see **Chapter 15**). In value disputes, moves to generate yet more scientific knowledge, to claim the science is settled, to communicate climate science ever more aggressively or to simply label your opponents as 'deniers', will not achieve much of lasting value. Rather what is needed is a clear articulation of the different values that are at stake in the dispute and then to engage in political processes to explore and reach decisions about what to do. Science provides no short-cut to this challenging and often messy task.

Given this perspective, *Contemporary Climate Change Debates: A Student Primer* approaches arguments about climate change primarily as value disputes that cannot be satisfactorily resolved simply by providing more scientific knowledge and asserting its truth more loudly or engaging in more vocal science communication. This is not to say that scientific knowledge about the climate is irrelevant. Of course not. Even less is it to suggest that I question the physical reality of climate change or the growing responsibility for many of these changes that humans must shoulder. I question neither of these facts, whatever Michael Mann may think is implied in the label 'climate contrarian'. Rather, it is to argue that climate change raises many complex and interlocking moral, ethical and political questions about the future, the answers to which lie beyond the reach of science. Examining these questions, and understanding how different scholars analyse and answer them in different ways, is a crucial learning experience for any student of climate change whether at high school, college or university. It opens the way for deeper appreciation of what is at stake with climate change. More broadly, it illuminates in fresh ways some of the enduring attributes of the human condition to which climate change draws our attention.

Questions of perspective, identity, value and prescription are central to many of the disagreements fostered by climate change. These same questions are central to the humanities and so *Contemporary Climate Change Debates* approaches climate change more from the humanities tradition than from that of the natural or social sciences (Kagan, 2009). As philosopher Richard Foley observes in his book *The Geography of Insight: The Sciences, The Humanities, How They Differ, Why They Matter*, 'individuals can have a better or worse understanding of prescriptive issues, and it can be an aim or inquiry [of the humanities] to increase this understanding. This cannot be the primary responsibility of scientific inquiries, however' (Foley, 2018: 29). *Contemporary Climate Change Debates* invites readers to increase their understanding of 'the prescriptive issues' that underlie 15 significant questions raised by the challenges of climate change.

Each of the chapters in the book answers one of these 15 questions, but it does so by presenting two *different* answers: one 'Yes', the other 'No'. These different answers are written by leading scholars to lay out the evidential and normative grounds—the descriptive and prescriptive bases—for these competing positions on each question. This is a different approach to writing a student textbook on climate change than one offering a systematic or synoptic account of the topic from a single author or one offering a primer of some climate research field. It is an approach intended to appeal to students studying climate change across the sciences, social sciences and humanities. In *Contemporary Climate Change Debates* you will hear different voices and arguments, shaped by different disciplinary and value commitments, whilst the book as a whole retains strong editorial cohesion. The authors are selected from 12 different countries: nearly half of the 36 contributors are from non-anglophone countries and there is a roughly equal number of male and female voices.

There is of course a large population of important climate change questions from which I selected these 15. My choices illustrate the range of cultural, economic, epistemic, ethical, legal, political, social and technological challenges raised by climate change. Following an opening framing question—'Is climate change the most important challenge of our times?' (**Chapter 1**)—the questions are grouped into four sections:

Part I: What do we need to know?

- Is the concept of 'tipping point' helpful for describing and communicating possible climate futures?
- Should individual extreme weather events be attributed to human agency?
- Does climate change drive violence, conflict and human migration?
- Can the social cost of carbon be calculated?

Part II: What should we do?

- Are carbon markets the best way to address climate change?
- Should future investments in energy technology be limited exclusively to renewables?
- Is it necessary to research solar climate engineering as a possible backstop technology?

Part III: On what grounds should we base our actions?

- Is emphasising consensus in climate science helpful for policymaking?
- Do rich people rather than rich countries bear the greatest responsibility for climate change?
- Is climate change a human rights violation?

Part IV: Who should be the agents of change?

- Does successful emissions reduction lie in the hands of non-state rather than state actors?
- Is legal adjudication essential for enforcing ambitious climate change policies?
- Does the 'Chinese model' of environmental governance demonstrate to the world how to govern the climate?
- Are social media making constructive climate policymaking harder?

I have used classroom debates in my higher education teaching in the UK for over a decade—with environmental science and geography students and with final year undergraduates and Master's students. For a '**wicked problem**' (see Glossary for emboldened concepts) like climate change, where there is no single correct position on *how* to deal with the challenge, nor *why* it should be dealt with this way, nor by *whom*, structured debates become effective learning devices for students. Stylised debating positions allow the interweaving of both descriptive ('this is known') and prescriptive ('this is right') arguments. In other words, through debate students learn not only about the state of academic knowledge on a topic, but also see how knowledge is politically and ethically sterile unless it is interpreted using strong normative reasoning. To paraphrase Hannah Arendt, it is necessary to pass judgement on the facts to be able to act politically in the world. Furthermore, through debate students learn that such reasoning often leads to disagreement. But they also learn that disagreement, far from being innately destructive, can at best be an opportunity for self-reflection and personal learning.

At a time when there is rising concern about the narrowness of students' educational experiences and their lack of exposure to people and/or views with which they disagree, this book is timely. With growing evidence of online echo chambers and strong social sorting (Mason, 2018), both feeding the rise of identity politics and populism in many societies, we owe our students a learning experience of climate change which exposes and explains the reasons for answering the challenging questions posed by climate change in different ways. *Contemporary Climate Change Debates: A Student Primer* is therefore a textbook which will help you develop your own well-informed position on these questions without being told what to think. You should be able to arrive at answers to complex questions, giving credible and reasonable accounts of their reasoning, without mere appeal to the authority of others or to calling down your own social identity. To quote Foley again, scholars and students alike 'should minimise the reliance on the opinions of others "floating in their brains" and should instead to the extent possible arrive at conclusions they are able to defend on their own' (Foley, 2018: 74).

Equally, it is important in a democracy to learn to disagree well, to realise that people with whom you disagree are not necessarily misguided, malicious or out to harm you. Their own life experience, education and moral or value commitments, might just mean that they see and interpret the world differently. Being able to recognise this, being able to engage in respectful debate and to learn from your antagonist, is the essence of learning. It helps break a deepening and polarising partisanship which is anathema for democratic deliberation.

The 15 questions debated in this book are presented in such a way as to contribute to this learning goal for students of climate change in the sciences, social sciences and humanities alike.

Cambridge, August 2019

Notes

1 See: www.theguardian.com/commentisfree/2015/dec/16/new-form-climate-denialism-dont-cele brate-yet-cop-21. Accessed 2 June 2019.
2 See: Climate change is finally on the agenda for 2020. But is it too late for debating? *Newsweek.* www.newsweek.com/climate-change-debate-too-late-1452759. Accessed 8 August 2019.

References

Callison, C. (2014) *How Climate Change Comes to Matter: The Communal Life of Facts*. Durham, NC: Duke University Press.

Foley, R. (2018) *The Geography of Insight: The Sciences, The Humanities, How They Differ, Why They Matter*. Oxford: Oxford University Press.

Howarth, C. C. and Sharman, A. G. (2015) Labeling opinions in the climate debate: a critical review. *WIREs Climate Change*. **6**(2): 239–254.

Hulme, M. (2009) *Why We Disagree About Climate Change: Understanding Controversy, Inaction and Opportunity*. Cambridge: Cambridge University Press.

Kagan, J. (2009) *The Three Cultures: Natural Sciences, Social Sciences and the Humanities in the 21st Century*. Cambridge: Cambridge University Press.

Mason, L. (2018) *Uncivil Agreement: How Politics Became Identity*. Chicago, IL: University of Chicago Press.

Rittel, H. W. J. and Webber, M. M. (1973) Dilemmas in a general theory of planning. *Policy Sciences*. **4**: 155–169.

1 Is climate change the most important challenge of our times?

Sarah E. Cornell and Aarti Gupta

Summary of the debate

This debate tackles the overarching question that frames the whole book, namely what is the significance of climate change for today's decision-making and political action in the world? **Sarah Cornell** argues that climate change *is* the most important challenge of our times because it now sets the global context for all the other social and ecological challenges the world faces. Unmitigated climate change puts at risk all other human development accomplishments and measures of wellbeing. Conversely, **Aarti Gupta** argues that only by addressing the more fundamental 'inequality crisis' of our times do we have even a hope of effectively addressing the climate crisis. In a context of persisting inequalities, characterising climate change as the most urgent challenge might even be potentially dangerous.

YES: Because climate change is changing everything (*Sarah E. Cornell*)

Introduction

In our times it has become clear that Earth's climate is changing as a consequence of people's activities. The state of scientific understanding of the climate system has been routinely deliberated and detailed in successive assessment reports of the Intergovernmental Panel on Climate Change (IPCC, 2014). These reports also assess the social and ecological impacts of a changing climate and the implications of climate adaptation and mitigation activities. The 'bare facts' are that anthropogenic (human-caused) greenhouse gases are currently accumulating in the atmosphere. The main greenhouse gas is carbon dioxide from fossil fuel use. Landscape changes such as deforestation, agriculture, wetland drainage and the large-scale restructuring of coastal zones are also important sources of carbon dioxide and of methane and nitrous oxide, the two other main greenhouse gases. Emissions of these gases are increasing year-by-year and their concentration in the atmosphere is rising (Hawkins, 2019, gives clear visualisations of the changes).

As the atmosphere's chemical composition changes, so too do its physical properties: the higher the concentration of radiatively active gases in the atmosphere, the stronger the greenhouse effect and the warmer Earth's surface, and ocean, becomes. These processes are well understood and can be predicted with high scientific confidence. At the heart of today's climate modelling are calculations of fluid motion and exchanges of heat between atmosphere and ocean. The ice sheets, land surface and life itself are also involved in shaping Earth's climate, so advances in climate research have progressively included representations of these dynamics too, resulting in the current generation of multi-component Earth system models (Simmons et al., 2016).

In short, climate change is a global and increasingly well understood phenomenon. But why should climate change be seen as more important than any other environmental, social or political issue of our times? I will present five arguments about the importance of climate change from the perspectives of different fields of knowledge, spanning different timeframes and scales.

The argument

Climate change changes the conditions for life on Earth

The importance of climate change can be seen when we set it in the context of our understanding of Earth as a living planet: climate changes are associated with fundamental shifts in how Earth functions. In the field of Earth system science, one of the most productive—and at times contentious—ideas of the last century is the Gaia hypothesis (discussed in Kleidon, 2004). This is the idea that life on Earth operates as an active and adaptive 'control system' for the whole planet, tending to maintain stability in the climatic and geochemical conditions that are conducive to life itself. Seen from a planetary perspective over the long time-scales of geological change and the large spatial scales of macroecology, life is part of a kind of co-evolutionary dance with the physical world. Whether or not the feedback interactions between life and its abiotic environment are ultimately self-regulating and generally stabilising, it is well accepted that life both shapes and is shaped by Earth's climatic conditions.

The IPCC's Fifth Assessment Report (2014) acknowledges that climate change amplifies the risk of fundamental shifts in all components of the climate system. The IPCC's remit is to be policy relevant, so much of its discussion of climate impacts has a rather human-centred and near-term emphasis. Yet current climate change is also intensifying Earth's largest scale and long-term processes, involving ice sheets, ocean circulation, the water cycle and the living biosphere. Climate change is altering the global cycles and complex biophysical and biogeochemical feedbacks of the elements of life—not just carbon, but also nitrogen, oxygen, sulphur and more (van de Waal et al., 2018).

Understanding these macro processes is still a major scientific challenge. It involves piecing together evidence found in observations of different components of the Earth system. This evidence includes long time-scale palaeo-records such as ice cores and sea-floor sediments, as well as near real-time observations. Thresholds of change are unlikely to be fully predictable (see **Chapter 2**), even though it is clear that the faster Earth warms and the hotter conditions get, the more abrupt some of the shifts are likely to be.

Climate change affects all living beings

Climate change is also a vitally important ecological challenge, if we take a perspective focused on the smaller scale of organisms and their interactions. Organisms differ in the ways they adapt and in their ability to adjust to different (and now often more changeable) conditions, so today's diverse ecosystems will respond to climate change in a variety of ways. But put bluntly, living beings will either adapt to changing conditions, move to places where conditions are more favourable—or die if they can neither adapt nor move.

The observed effects of climate on ecosystems are increasingly well-documented scientifically (see a collation of recent research in PLOS & Atkins, 2016). Climate-driven ecosystem impacts are summarised in global assessment reports (e.g. IPBES, 2019; IPCC, 2014) and in studies supporting the UN Convention on Biological Diversity (www.cbd.int/climate). Climate change is already driving wholesale shifts in ecological regimes. Organisms are migrating to higher latitudes or higher altitudes, where temperatures are cooler. Range shifts change the assemblages of organisms that make up ecosystems and thus change the ways that ecosystems function. Range shifts of pests and diseases are a particular concern because they can drive very rapid disruptions. Effects of climate change on one species can also ripple through the food web.

Figure 1.1 shows how phenological changes, or shifts in the timing of life cycle events, can differ for different species. For example, caterpillar populations peak earlier than they used to because insects tend to respond rapidly to warmer conditions. Chicks are not hatching earlier however, because for birds egg laying is prompted by springtime day length more than by temperature. This means that fledgling chicks now have their highest food demands when their preferred food source is no longer abundant. Will species be able to adapt—or evolve—fast enough to keep up with such complex changes in their environment?

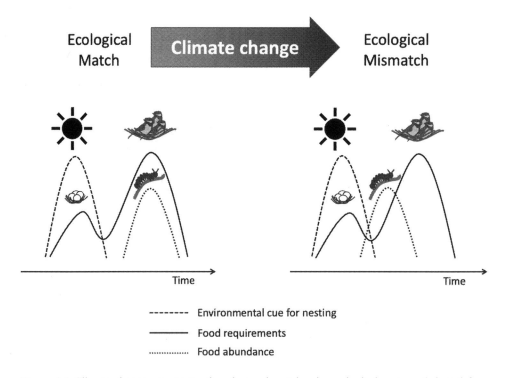

Figure 1.1 Climate change can create phenology mismatches in ecological systems (adapted from Stenseth and Mysterud, PNAS 99(21): 13379–13381; 2002; © 2002 National Academy of Sciences, USA). Great tits are one well-documented example: the environmental cues that trigger egg laying are no longer synchronised with the prevailing environmental conditions when chicks are reared and when birds' energetic demands are the highest

Some species and ecosystems (coral reefs and polar ecosystems, for instance) are conspicuously climate sensitive and are often described as 'highly vulnerable'. Coral bleaching and ice-melting at today's rates are severe threats and they are largely irreversible. But so too are other changes that are less visible to our human eyes, yet are no less important from an ecological viewpoint. Even though much of science and policy places life into categories, typologies and classifications, climate change is driving interconnected shifts in ecosystems and altering the diversity of life at all scales. Also, ecosystems are not hermetically sealed units. Whenever one ecosystem is 'lost', the ecological connections that are unravelled then start their re-weaving of the whole web of life of which our own species, *Homo sapiens*, is part. We may not need to face a mass extinction event to discover that we are maladapted to newly emerging ecological conditions.

Climate change has impacts on people

We should also look at today's climate change from the perspective of people's needs and wellbeing. While some climatic changes might actually make life easier for some, coping with current changes is already a major challenge for many people. Many studies document how climate change is a significant contributor both to present-day impacts and, likely, to future risks to people's survival and wellbeing all around the world. Changing temperatures, rainfall, storminess and other climate-related hazards have immediate effects on people's health, safety, welfare and livelihoods. The IPCC (2014) frames the key climate risks in terms of reduced food and water security, adverse effects on health and livelihoods and systemic risks linked to the breakdown of critical infrastructures. It also recognises the losses of ecosystem goods, functions and services because of disruptions to ecosystems and biodiversity.

These changes span across world regions, so the economic costs of current climate impacts are enormous—and rising. The World Bank's Shockwaves report (Hallegatte et al., 2016) focuses on climate change and poverty reduction, but its pervasive message is that issues of poverty, prosperity, productivity and equity cannot be isolated in today's highly globalised and interdependent world. Vulnerability to climate change is one among many systemically linked vulnerabilities. Importantly, climate change is one issue where good, targeted action will be a cost-effective way to steer towards many other societal goals.

Climate change is tightly linked to societal change

History shows that when climate changes, options also change for what societies can do. That is not to say that climate controls or determines social conditions and development trajectories. But seen from a 'longue durée' perspective of history and archaeology, climate change has always and everywhere been an important factor in the existential challenges and opportunities faced by the world's societies (Lane et al., 2018 provide a multidisciplinary reflection).

The traces of the profound importance of climate change can be uncovered in the emergence, sustainability and 'collapse' of societies. Climatic influences are evident in the location and layout of cities, in trade routes, arenas of war and other patterns of settlement and movement (see **Chapter 4**). Ancient ruins and many of today's technologies alike can be read as the enduring marks of societies' efforts to deal with sea-level rise and retreat, booms and busts of natural resource flows, the crises of climate-related plagues and weather extremes and the sometimes equally disruptive challenges of slower but cumulative environmental change.

We can expect the same of current and future global warming. Many physical processes of climate change are now locked in for centuries to come, even if strong action is taken to reduce greenhouse gas emissions. In coming years, societies worldwide will have to deal with altered ecosystems, food and hydrological systems, infrastructure, landscapes and livelihoods. The IPCC (2014) observes that social systems may be fundamentally transformed not just by climate impacts, but also by climate mitigation and adaptation actions. Mitigation pathways to stabilise climate (whether at 1.5°C, 2°C or higher) will require anthropogenic emissions of GHGs to be drawn down to zero or below; and achieving the lower temperature targets will require rapid and wide-ranging social, technological and economic changes. Whether today's societies can pursue resilient and sustainable pathways as they navigate unprecedented environmental conditions is a wide open question.

Climate change is our challenge, in our times

Today's technologically equipped societies have the scientific capacity to detect and attribute human-caused climate disruption and (crucially) to predict many aspects of the future climate with high confidence. Climate science—actually a field combining multiple areas of expertise—has provided a sophisticated understanding of the causes, effects, impacts and risks of climate change. It confirms that today's climate change differs in some key ways from changes in the past. In particular, it shows that people's collective activities are driving global heating and that there is scope for societies to take action to limit the rate and magnitude of climate change. New challenges for scientists and citizens alike come with this current power to anticipate and attribute climate change. Climatic changes of today's magnitude are not just of scientific interest; it is also important for wider society to know about them. Initiatives like NASA's Climate Apps (NASA, 2019) translate Earth observations and scientific insights into publicly available appealing and informative maps and charts of environmental conditions. Since 1990, the IPCC's successive scientific assessment reports have intensified their messages to policymakers about the need to prioritise climate action. And the tone of concern remains, despite the attenuation of their messages that comes from the need to achieve international political consensus on the wording.

Industrialised technology-dependent societies have an especially important role to play in responding to climate change. Allocating historic responsibility may be fraught with ethical challenges (see **Chapter 10**), but regardless of past contributions to the climate problem the people in these societies today contribute by far the most to the drivers of climate change. And yet shrinking this large climate footprint is difficult, even seemingly unimaginable to some people. At present, many responses to changing climate conditions are actually fossil energy-intensive (for instance, installing more electrical air conditioners for thermal comfort), creating problematic reinforcing feedbacks where efforts to cope with hotter conditions contribute to more global heating. Another, often problematic, example is carbon offsetting, where rich countries, companies and individuals can continue their high-emission activities by buying offsets or credits to support carbon-reducing projects elsewhere. These schemes do not actually reduce overall emissions, and some kinds of offset schemes—such as monoculture forest plantations—are not always good for the environment or the local communities involved (again, these shortcomings are documented in successive IPCC assessment reports; e.g. IPCC, 2014). Even though technology enables many current societies to be buffered against the impacts of climatic changes, too often it simply externalises one society's impacts to other places. In this way, maladaptation and mismanagement of mitigation actions can aggravate today's climate change problems.

The issue of whether climate change is the *most* important challenge of our times is obviously value-loaded and so needs careful reflection. Climate change is increasingly recognised as both important and urgent, not just in academia but also in growing areas of business, policy and society at large. And there is growing awareness that the cost of failing to reduce emissions is rising, whatever measures of wellbeing are used to define 'cost'. Some have argued that other issues are more important than climate and should be prioritised for investment and action now (for example, this is Danish economist Bjørn Lomborg's position[1]). However, this perspective implies that society's issues are all separate, distinct and rankable, rather than all playing out together in the context of a changing climate.

The globalised and interconnected cultural and economic systems of our times mean that climate risks are now changing everywhere, at all scales and in all timeframes. It is difficult to defend the idea that particular sectors or places are vulnerable to climate change while others are not. (Of course, particular groups of people may actually be *made* vulnerable by political choices, because such decisions leading to exclusion, marginalisation and inequity also happen regardless of climate conditions.) Short-term cost-effectiveness considerations no longer represent simple trade-offs with long-term societal investments for sustainability. There is growing consensus and rising concern that the projected future dangers of climate change are already intensifying and are putting other human development accomplishments at risk. In short, in these challenging times, climate change really needs to be taken into consideration in all other decision-making contexts.

Conclusion

Is climate change the most important challenge of our times? This is not the kind of question that can be assessed, answered and explained through a reliance on general scientific laws. The question is an empirical one, in the sense of being aware of our experiences of these times. The nineteenth century educator and philosopher William James argued that 'an idea is made true by events'. So we might argue from this perspective: the idea that climate change is a pressing challenge has now been articulated from many different perspectives, just five of which I have flagged here. The worldwide stream of observations, experiences and events is making it empirically true that climate change *is* the most important challenge we face.

But *should* climate change be the most important challenge of our times? This is a normative question, opening the issue up to debate, choices and action. It is definitely now *our* challenge, whether we are acting as individuals, societies, scholars, policymakers or market actors—and of course most people have several such roles and identities at any given time in everyday life. People may face a great deal of dissonance among these various roles when making assessments of whether climate change is important enough to shape action. The task of both personal and collective sense-making about climate change involves questioning and weighing up what we think we know against experiences and observations, wherever they may come from. Even if we may not feel challenged ourselves at this time and in our locality, it would be a perverse kind of deliberate blindness to ignore both the realities elsewhere and the complex global tangles that connect those realities to us. And we should remain aware that denial and the suppression of different perspectives have long been tools of the powerful in framing important issues as unimportant.

Climate change demands attention to local and global, ecological and societal, analytical and ethical perspectives—and more—all at once. Rigorous relevant scholarship is needed more than ever, because the multidisciplinary perspectives and predictive powers of science in our times connect with issues that are well beyond the traditional scope of academia alone. The English word *science* is often taken to mean the state of knowledge, particularly about the physical and technical worlds. In a rapidly changing world, we need to shift to a more active process of 'knowledging'—as in *Wissenschaft*, the German word for science—which holds this dynamic sense of knowledge making and sharing. Fundamentally, climate change is playing out as the ultimate knowledge-and-action challenge. And that involves us all.

NO: Because we cannot address climate change without addressing inequality (*Aarti Gupta*)

Introduction

Climate change is one of the most complex challenges facing human societies in the twenty-first century. Yet what *kind* of challenge is it? In this essay, I argue that climate change is, at its core, a political challenge. If so, the most fundamental questions of politics (*who* gets what and *why*) need to be addressed head-on in considering solutions to this pressing challenge. This is particularly true in light of the increasingly dominant claim that climate change is a looming *collective* challenge, an urgent crisis threatening 'all of humanity'. But humanity in the twenty-first century is far from a homogenous mass, where individuals have equal impact on—or access to—resources and life opportunities. To the contrary, structural inequalities are growing between and within societies (UN-DESA, 2016).

This essay argues, therefore, that *growing inequality* is the greatest challenge facing humanity in the twenty-first century. Without explicitly acknowledging and addressing multiple forms of inequality, we cannot hope to forge an effective—let alone fair—global response to climate change. As Myles Allen, one of the lead authors of the IPCC's 2018 Special Report on 1.5°C points out, 'one of the most insidious myths about climate change is the pretence that we are all in it together'.[2] Taking this line of argument further, I posit here that in the face of the persisting structural inequalities within which the climate problem is diagnosed and responses are forged, it can even be dangerous to characterise climate change as the most important challenge facing us.

Why dangerous? Because a climate crisis or emergency framing can be used to justify the taking of extraordinary measures. These *could* be emancipatory, with climate change serving as a clarion call for fundamentally changing course and reorganising the dominant (and unequal) modes of production and consumption that fuel the current climate crisis. Yet in a world of obscenely unequal access to resources, wealth, power and economic and political opportunity, extraordinary measures are more likely to mean the opposite. They could signal the setting aside of hard-fought gains for equal rights and access to economic and political opportunity by the historically marginalised within and across societies. And this could be done in the name of climate actions deemed necessary to meet a challenge ostensibly facing us all, but in which much of humanity is not consulted in devising so-called emergency solutions.

The above considerations underpin the 'No' position taken here to the question posed in the title of this chapter. I consider three potential risks of casting climate change as the most compelling challenge facing us today: (a) the risk of *setting aside distributive justice considerations*, as hindrances that impede an urgently needed, swift and effective collective response to climate change; (b) the risk of *setting aside democratic decision-making procedures*, in favour of authoritarian or technocratic approaches seen as essential to forging a timely and effective response; and (c) the risk of *considering reliance on unilaterally deployable, highly controversial and risky techno-fixes*, such as **solar climate engineering**. I address each of these risks below.

Climate change and inequality: twin challenges, inextricably linked

A crisis framing around climate change does not usually start from identifying the multi-faceted, structural inequalities underpinning the climate crisis (but see Klein, 2014). It starts from diagnosing the nature of the climate problem primarily in terms of its atmospheric and biogeochemical impacts and consequences, which are seen as potentially catastrophic for all and hence urgent to address. Such analyses see climate change as threatening all other human accomplishments and as a threat multiplier, or '**wicked problem**' (see Hulme, 2009: 334–337, for a nuanced discussion of the merits and limitations of this latter notion).

In such a framing, a primary challenge becomes how to most efficiently address the direct cause of climate change: emissions of greenhouse gases. Understood thus, proposed solutions to the problem of climate change focus on how to best harness human ingenuity, science and technology, corporate power and state authority to reduce greenhouse gas emissions and simultaneously adapt to the worst ravages of climate change, so as to get ourselves out of a collective and looming mess. The crucial and less explicitly addressed challenge, however, is how to do the above in the context of the structural inequalities that underlie both the climate crisis and responses to it (Klein, 2014; Paterson, 2018). This relates fundamentally to diverse views on what should constitute 'climate policy' in the first instance. As a slew of progressive political analysts have suggested, any problem diagnosis of climate change needs to extend well beyond greenhouse gas emissions and their management (e.g. Holmberg, 2017). As Becky Kelley, president of the Washington Environmental Council, explains, 'viewing climate change as an environmental [issue] is way too limited. Climate policy is not environmental policy. *It is everything policy*. It is transformational, societal, policy [concerned with] economics and social justice' (emphasis mine).[3]

In what way is climate policy 'everything policy'? Critical political economy perspectives have long located the drivers of climate change in capitalist (unequal) structures of production and consumption (e.g. Paterson, 2018). From such a perspective, effective responses necessarily need to target such underlying drivers, including structural inequalities. At the very least, envisioned solutions to the climate crisis should encompass, *inter alia*, a living wage, access to education and health care and steady and gainful employment (UN-DESA, 2016; Holmberg, 2017, 2016). It is in this sense that climate policy is 'everything policy', a notion that also underpins growing policy and scholarly debate on the need for a 'just transition' to a low carbon economy (e.g. Newell and Mulvaney, 2013).

Such an expansive framing of the climate crisis is exemplified, for example, by the much-debated Green New Deal recently advocated by Democratic congresswoman Alexandria Ocasio-Cortez and her supporters in the United States Congress. Here, combating climate change and combating structural inequalities and securing more equal economic opportunity are seen as fundamentally intertwined. This was also the resounding message of the *gilets jeunes*—Yellow Vest—protests sweeping France during 2018–19, the message that inequality and climate change are inextricably linked. One problem cannot be effectively—let alone fairly—addressed without addressing the other. This holds within national contexts, as the examples reveal, but also globally (Rao and Min, 2018, UN-DESA, 2016).

And herein, I argue, lies the fundamental challenge that the looming climate crisis poses. If we accept the link between climate change and inequality, then what institutional and governance mechanisms do we have at hand to tackle it? How likely is it that we will want to do so? Are there historical precedents for overcoming large-scale inequalities while meeting intertwined global challenges—whether ecological, political, economic, or threats to global peace and security? Naomi Klein takes on the task of examining these opportunities and possibilities in her provocatively titled book about climate change—*This Changes Everything* (Klein, 2014)—a trenchant critique of the links between the climate crisis and capitalism. For example, she documents that for the politically and economically marginalised it is economic disadvantage that has historically been the hardest to counter. Using slavery as an example (but this holds also for the women's rights and civil rights movements), history shows that securing equal legal and political rights for the marginalised is easier than reducing unequal economic opportunity.

Yet it is this latter imperative that, I posit here, makes climate change such an intractable challenge—*not* primarily because its biogeophysical dimensions are worrisome, complex and fraught with uncertainties, nor even because it is a 'wicked' multisectoral issue. It is intractable because of the extreme political difficulty of tackling the structural inequalities underpinning the climate crisis in the first instance. If, however, inequality and climate change were to be recognised as inextricably linked, with political acknowledgement that the two need to be addressed together, climate change *could* serve as a powerful rallying cry for fundamental transformations of currently unequal global systems of resource access and use and unequal access to power and opportunity (Klein, 2014; see also Rao and Min, 2018). Yet, it remains to be seen whether such a radically different political agenda will find sufficient support.

It should be noted that vulnerabilities of the poorest and the most marginalised to the adverse *effects* of climate change do not go ignored in an 'urgency' framing of the climate problem. Indeed, such vulnerabilities are frequently emphasised when framing climate change as the most pressing challenge facing humanity. As part of such a framing, mainstream analyses of the climate crisis routinely evoke unequal impacts, capacities and vulnerabilities of the poor and the marginalised to the worst ravages of climate change. What is much less acknowledged however, including in policy responses, are the historical trajectories of colonialism and extractivism between and within states that have fuelled cycles of poverty and environmental degradation. These have allowed the rich (whether states or individuals; see **Chapter 10**) the capacity to pollute the environments wherein the poor primarily reside, as only one example of these relationships (see

Holmberg, 2017, for a more detailed analysis). Thus, even as *inequality of impacts* (primarily biophysical, but increasingly also social impacts) are widely evoked in mainstream discussions of climate change, there is little acknowledgement of inequality as a significant *driver* of climate change (UN-DESA, 2016).

Such considerations will increasingly come to the fore in light of the 2015 Paris Agreement's aspirational and ambitious temperature target of keeping average increase in global temperatures to below 2°C, while pursuing efforts to limit the warming to 1.5°C. It is widely acknowledged that temperature increases higher than 1.5°C above pre-industrial levels will disproportionately impact on the poorest, most vulnerable and most marginalised in the world. But an equally valid question, one not posed often enough, is: who will bear the burden of striving for 1.5°C, particularly in light of vastly unequal emission trajectories (e.g. Chancel and Piketty, 2015)? Will the impacts of addressing climate change also come to be laid disproportionately at the feet of the most vulnerable? (How) do these dynamics change when climate change is cast as the most important threat facing humanity? In portraying climate change in such a way, what gets de-emphasised? And what comes to the fore?

My aim here is not to offer one specific take (mine) on the questions asked above. Instead, I suggest that it is the *posing of these questions* that is of singular importance. It is through asking such questions that we can begin to ascertain the diverse perspectives on the collective urgency of the climate challenge and suitable responses to it. Doing so makes apparent that the answers will be very different. They will depend upon problem diagnosis, worldview and positionality in terms of vulnerabilities, capacities and exposure to harm (see Hulme, 2009 for an early and extensive discussion of how we collectively talk about climate change). Thus, any notion that we can all agree that climate change is self-evidently the most important challenge of our times needs to keep this reality in mind.

In the absence of such awareness, we can too quickly embrace 'solution' pathways that in the long run exacerbate, rather than alleviate, the climate crisis for all. In the remainder of this essay, I consider two potential further risks in framing climate change as the most important challenge facing us today. First, that it might privilege more authoritarian or technocratic decision-making processes, rather than messy, slow and unpredictable democratic ones; and second, that it might privilege the use of speculative, high-risk and unilaterally deployable climate engineering options as the most 'viable' way out of a climate crisis.

Democracy in an age of climate: the tyranny of urgency?

The growing urgency around climate change has been accompanied by intensifying debate about the merits of democratic approaches to addressing this multifaceted challenge (e.g. Stehr, 2016). This debate is salient not only in national contexts, but also in global settings where the question of democracy in the Anthropocene looms large. As Amanda Machin has suggested, the notion of humanity as a disruptive geological force points to an 'irresolvable political paradox'. This is reflected in the fact that *boundaries* are inherent in the notion of a political demos such as the state, even as an Anthropos-framing implies a homogenous humanity *transcending* political boundaries. As she explains, this paradox calls for a 'lively democratic politics in which the demos is always prompted to reimagine itself and ask, who are "we" in the Anthropocene?' (Machin, 2019: 1). When assessing the implications of framing climate change as the most important challenge currently facing 'us', this is an important first-order question that needs to be asked.

The issue of democracy has come to the fore because some prominent (and worried) climate scientists, among others, have contemplated setting aside democratic decision-making procedures in the interest of taking quick and effective policy action in the long-term interest of all. Expressing the pessimistic view that democratic systems will never deliver effective action in time, their underlying rationale is that the intractable nature of the climate problem requires such an anti-democratic move. For example, environmental philosopher Dale Jamieson suggests that climate change

> is the largest collective action problem that humanity has ever faced...[but] we have not designed the political institutions that are conducive to solving them ... Sadly, it is not entirely clear that democracy is up to the challenge of climate change.
>
> (Jamieson, 2014, quoted in Stehr, 2016: 2)

This sentiment has been echoed by other prominent climate scientists and thinkers, including James Hansen and James Lovelock. A common refrain is to compare climate change to war, with the argument that in such exceptional circumstances democracy is inconvenient (Stehr, 2016 contains an extensive overview of these diverse perspectives; see also Brown, 2014).

Yet does an effective response to climate change require setting aside democratic procedures? Many scholars and political commentators emphatically argue the opposite. In their defence of democracy, they note the potential risks inherent in what Strobe Talbott, ex-President of the Brookings Institution in the United States, refers to as the 'tyranny of the urgent'.[4] In this political move, one problem comes to be elevated as being the most urgent, above all others, thus requiring immediate attention even if this means setting aside normal decision-making procedures.

Rather than evoking the benefits of authoritarianism (see **Chapter 14**), climate scientists questioning the merits of democracy are more likely to (implicitly) advocate for reliance on *technocracy* instead; i.e. relying on scientists and scientific input in devising solutions, rather than leaving these to emerge from messy democratic political processes. Yet, as Stehr (2016) and others have noted, this elevating of experts and expertise to the pinnacles of power, in the guise of delivering technocratic solutions, is counter to the very notion of climate change as a fundamentally *political* challenge.

The contested political nature of climate change can be strikingly illustrated by the controversial topic of climate deniers (in the USA and elsewhere) and their view of the role of democracy in public life. In his discussion of the climate crisis and democracy, Brown (2014) shows the current gulf between, on the one hand, climate scientists advocating for technocracy and, on the other, populists (or climate deniers) categorically rejecting climate science as yet another source of established power. In analysing these dynamics and their implications for democracy, Brown (2014) advances the provocative premise that

> rejectionism [of climate science] is not simply an unwillingness to face the 'inconvenient truth', but a political reaction against those *who would use truth to eliminate*

politics. In this respect, those who reject mainstream climate science may inadvertently promote a more democratic approach to climate science and policy.

(Brown, 2014: 141, italics added)

This provocative claim merits much more scrutiny (see also **Chapter 15**). But the general point is that while climate science has an important role in political debates, it cannot, as Brown states, 'determine which policies best represent the needs and values of diverse human communities around the globe' (p. 141). This resonates with the notion that, ultimately, climate change is an idea and an opportunity to imagine diverse futures and how we want to live (Hulme, 2009). The challenge remains how to democratically *and* collectively imagine these myriad desired futures—and move towards the realisation of them.

Climate change and looming techno-fixes: the hubris of climate engineering

There is another risk inherent in an urgency framing of climate change. This is the risk that claiming that climate change is the most important challenge facing us today can be used to justify research into, and potential future unilateral deployment of, speculative, untested and controversial climate engineering technologies. Advocates of such technologies increasingly suggest that solar climate engineering—deliberately reflecting some amount of incoming solar radiation back into space through technological means—might be the only 'viable' way out of an otherwise intractable collective climate disaster (see **Chapter 8**). They further argue that climate engineering is a technology essential to contemplate for the sake of the most vulnerable, arguing that those least responsible and most vulnerable are *owed* solutions like climate engineering in the face of a looming crisis not of their making.

Such an 'equity' rationale to pursue climate engineering, advanced by some of its advocates, has been heavily criticised because it inverts justice concerns relating to these technologies. It frames the poor and marginalised as passive, grateful recipients of what are in reality speculative, controversial and risky large-scale interventions. Because of their historical responsibility for the climate crisis, the rich are portrayed as global risk managers *for* the poor (Flegal and Gupta, 2018), with this responsibility now being exercised by offering solutions such as climate engineering. This is a highly problematic move. Such a framing ignores the privileged position of the technology's advocates in providing—and profiting from—such untested and risky options. Such technological adventurism will also likely exacerbate inequalities and worsen, rather than mitigate, the long-term consequences of the climate crisis (e.g. Preston, 2016).

Conclusion

In this essay I have argued that casting climate change as *the* central challenge facing humanity risks creating a post-political, post-equity and post-democratic world. In such a world, in order to act quickly against this 'global' threat to humanity writ large, long-standing distributional justice concerns are marginalised, democratic procedures set aside and/or controversial and high-risk techno-fixes relied upon. Yet, in the context of the twenty-first century's extreme and growing inequality, who will pay the price of such outcomes? All these are fundamentally political moves, even if they are often framed as *setting aside politics* in order to take decisive and effective action against climate change. Yet herein lies the central error. In order to effectively address the compelling, complex and multidimensional problem of climate change, we need *more* democratic and just governance, not less. It is precisely securing *this* goal—just and democratic govern-ance—that is the most important challenge of our time.

Further reading

Chancel, L. and Piketty, T. (2015) *Carbon and Inequality.* Paris: Paris School of Economic Study.

This study presents the evolution of the global distribution of greenhouse gas emissions amongst the world's citizens from 1998 to 2013 and examines different strategies for a global climate adaptation fund based on efforts shared among high emitting individuals rather than between high-income coun-tries. Results depend not only on within-country inequalities, but also on consumption-based GHG emissions of different countries. Available at: http://piketty.pse.ens.fr/files/ChancelPiketty2015.pdf.

Connors, S. and Pidcock, R. (2018) 'Frequently Asked Questions', in: *Global Warming of 1.5°C. An IPCC Special Report on the impacts of global warming of 1.5°C above pre-industrial levels and related global greenhouse gas emission pathways, in the context of strengthening the global response to the threat of climate change, sustainable development, and efforts to eradicate poverty.* Geneva, Switzerland: World Meteorological Organisation. Available at: www.ipcc.ch/sr15/faq

This web resource offers a useful set of short and clear answers to key questions about climate change, about the consequences of different levels of warming and about the likelihood of the world's future development path delivering these levels. A good primer for the whole book, it is also useful in thinking through how states and non-state actors can deliver on different cli-mate goals.

Newell, P. (2019) *Climate and Development: A Tale of Two Crises*. University of Sussex.

This Institute of Development Studies lecture from the University of Sussex (available on YouTube) looks at the intertwined histories of development and climate change and argues that only a very dif-ferent approach to development can help to address the climate crisis we currently face. Available at: www.youtube.com/watch?v=2a8xOgl1K0E.

Schmidt, G. and Wolfe, J. (2009) *Climate Change: Picturing the Science*. New York: W. W. Norton & Co.

This book shows how human-caused climate impacts are already observable and how societies around the world are responding to climate challenges. It brings together many technical aspects of climate change science into the scope of people's lived experience, making it a rich resource for the wider understanding of climate science.

Schneider, S.H. (1996) *Laboratory Earth: The Planetary Gamble We Can't Afford To Lose*. London: Weidenfeld & Nicolson.

This book, written 20 years ago, explains many of the concepts of Earth system science and describes the co-evolution of life and the climate system over long timeframes. Schneider also wrote enga-gingly and informatively about the social and political challenges of climate change. Techniques for global change modelling and Earth observation have advanced greatly since then, but the book's main insights are robust and are worth reading.

World Economic Forum (2019) *Global Agenda Reports*. Cologny, Geneva, Switzerland.

Global Agenda reports offer diverse perspectives on the most pressing challenges facing the world today. See, for example: *Why income inequality is bad for the climate*. Available at www.weforum.org/agenda/2019/01/income-inequality-is-bad-climate-change-action/ and *Emissions inequality: there is a gulf between global rich and poor*. Available at www.weforum.org/agenda/2019/04/emissions-inequality-there-is-a-gulf-between-global-rich-and-poor/.

Follow-up questions for use in student classes

1. Can we claim objectively that climate change *is* the most important challenge facing humanity, or is such a claim always a subjective value-judgement?
2. Reports about options for climate adaptation (including IPCC, 2014; e.g. Table 2.3) often include statements such as, 'poor people and women are most at risk from climate change'. What does this framing imply about people's adaptation options and about the assumptions embedded in such reports about today's and future societies?
3. Given the likely societal costs and disruptions of overhauling worldwide energy, food and transport systems to mitigate climate change, is being concerned about climate change actually an 'anti-human' outlook?
4. Do you agree that climate change and inequality are inextricably linked? Why or why not?
5. Should democracy be abandoned if it prevents quick action on climate change? And what notion of democracy would you be abandoning?

Notes

1 See: www.lomborg.com/. Accessed 24 July 2019.
2 Available at: https://theconversation.com/why-protesters-should-be-wary-of-12-years-to-climate-break down-rhetoric-115489?utm_source=twitter&utm_medium=twitterbutton. Accessed 25 June 2019.
3 Available at: https://thinkprogress.org/washington-carbon-tax-campaign-7ce90a306e7f/#.1nsxkugbj. Accessed 25 June 2019.
4 Available at: www.ft.com/content/c14645d2-a8f8-11dd-a19a-000077b07658. Accessed 25 June 2019.

References

Brown, M.B. (2014) Climate science, populism and the democracy of rejection. In D.A. Crow and M.T. Boykoff, eds. *Culture, Politics and Climate Change: How Information Shapes our Collective Futures*. Abingdon: Routledge. pp. 129–145.

Chancel, L. and Piketty, T. (2015) *Carbon and Inequality*. Paris: Paris School of Economics Study. Available at: http://piketty.pse.ens.fr/files/ChancelPiketty2015.pdf

Flegal, J. and Gupta, A. (2018) Evoking equity as a rationale for solar geoengineering research? Scrutinizing emerging expert visions of equity. *International Environmental Agreements: Politics, Law and Economics*. **18**(1): 45–61.

Hallegatte, S., Bangalore, M., Bonzanigo, L., Fay, M., Kane, T., Narloch, U., Rozenberg, J., Treguer, D. and Vogt-Schilb, A. (2016) *Shock Waves: Managing the Impacts of Climate Change on Poverty*. Washington, DC: Climate Change and Development Series, World Bank. Available at https://openknowledge.worldbank.org/handle/10986/22787

Hawkins, E. (2019) *Climate Spirals*. [online] Climate Lab Book – open climate science. Accessed 25 June 2019. www.climate-lab-book.ac.uk/spirals

Holmberg, S.R. (2017) *Boiling Points: The Inextricable Links between Inequality and Climate Change*. New York: The Roosevelt Institute. Available at https://rooseveltinstitute.org/wp-content/uploads/2017/05/SHolmberg_ClimateReport.pdf

Hulme, M. (2009) *Why We Disagree about Climate Change: Understanding Controversy, Inaction and Opportunity*. Cambridge: Cambridge University Press.

IPBES. (2019) *IPBES Global Assessment Summary for Policymakers*. Intergovernmental Science-Policy Platform for Biodiversity and Ecosystem Services. Bonn, Germany. Available at: www.ipbes.net/sites/default/files/downloads/spm_unedited_advance_for_posting_htn.pdf

IPCC (2014) *Climate Change 2014: Synthesis Report. Contribution of Working Groups I, II and III to the Fifth Assessment Report of the Intergovernmental Panel on Climate Change*. Geneva, Switzerland: WMO.

Kleidon, A. (2004) Beyond Gaia: thermodynamics of life and Earth system functioning. *Climatic Change*. **66**(3): 271–319.

Klein, N. (2014) *This Changes Everything: Capitalism vs. the Climate*. New York: Simon & Schuster.

Lane, M., Sörlin, S., Socolow, R.H. and McNeill, J. eds. (2018) Responding to climate change: studies in intellectual, political, and lived history. *Climatic Change* (Special Issue). **151**(1): 1–78.

Machin, A. (2019) Agony and the anthropos: democracy and boundaries in the Anthropocene. *Nature and Culture*. **14**(1): 1–16.

NASA (2019) *Climate mobile apps*. NASA Jet Propulsion Laboratory. Accessed 25 June 2019. https://climate.nasa.gov/earth-apps

Newell, P. and Mulvaney, D. (2013) The political economy of the 'just transition'. *The Geographical Journal*. **179**(2): 132–140.

Paterson, M. (2018) Political economies of climate change. *WIREs Climate Change*. **9**: e506. 10.1002/wcc.506.

PLOS Ecology Community and Atkins, J. (2016) Ecological impacts of climate change collection, 2015–2016. *PLOS* Blogs. Accessed 25 June 2019. https://blogs.plos.org/ecology/2016/08/01/ecological-impacts-of-climate-change-collection-2015-2016

Preston, C., ed (2016) *Climate Justice and Geoengineering: Ethics and Policy in the Atmospheric Anthropocene*. London: Rowman and Littlefield.

Rao, N.D. and Min, J. (2018) Less global inequality can improve climate outcomes. *WIREs Climate Change*. **9**: e513. 10.1002/wcc.513.

Simmons, A., Fellous, J.L., Ramaswamy, V., Trenberth, K., Asrar, G., Balmaseda, M., Burrows, J.P., Ciais, P., Drinkwater, M., Friedlingstein, P., Gobron, N., Guilyardi, E., Halpern, D., Heimann, M., Johannessen, J., Levelt, P.F., Lopez-Baeza, E., Penner, J., Scholes, R. and Shepherd, T. (2016) Observation and integrated Earth-system science: a roadmap for 2016–2025. *Advances in Space Research*. **57**(10): 2037–2103.

Stehr, N. (2016) Exceptional circumstances: does climate change trump democracy? *Issues in Science and Technology*. **23**(2-Winter). Available at https://issues.org/exceptional-circumstances-does-climate-change-trump-democracy/

Stenseth, N.C. and Mysterud, A. (2002) Climate, changing phenology, and other life history traits: nonlinearity and match–mismatch to the environment. *Proceedings of the National Academy of Science*. **99**(21): 13379–13381.

UN-DESA [United Nations Department of Economic and Social Affairs]. (2016) *The Nexus Between Climate Change and Inequalities*. UN-DESA Policy Brief No.45. New York.

van de Waal, D.B., Elser, J.J., Martiny, A.C., Sterner, R.W. and Cotner, J.B., eds. (2018) *Progress in Ecological Stoichiometry*. Lausanne: Frontiers Media.

Part I

What do we need to know?

2   Is the concept of 'tipping point' helpful for describing and communicating possible climate futures?

Michel Crucifix and James Annan

Summary of the debate

The concept and terminology of tipping points has been widely used to invoke the danger of passing thresholds of irreversible and/or abrupt change in the near immediate future. But how helpful is this metaphor for climate science and for climate change communication? **Michel Crucifix** argues that the tipping point concept may have some limitations as a description of the mathematical behaviour of Earth system models trying to simulate the world's climate. But its use can alert decision-makers to the possibility of some rapid and/or serious changes in the climate system which need attention as part of responsible and accountable policymaking. In contrast, **James Annan** argues that the concept risks exaggerating the immediacy and severity of climate change and offers a false prospectus of there being a 'cliff edge' within the climate system. Climate change is primarily a problem of incremental cumulative harm and the concept of tipping point offers a false emphasis on abruptness and harms the public understanding of climate science.

YES: It draws attention to the possibility of inadequate response to non-incremental change (*Michel Crucifix*)

Introduction

To quantify the effects of emissions of carbon dioxide (CO_2) and other greenhouse gases, scientists and policy advisers commonly refer to the concept of **climate sensitivity**. This concept is based on the reasoning that a change in CO_2 concentration perturbs the radiative balance of the planet. Radiative equilibrium is then restored by a change in temperature. It is generally assumed that this temperature response is roughly proportional to the forcing. If the climate sensitivity is, say, 3°C (a value generally considered as plausible), then a doubling of CO_2 concentration will eventually cause a global average surface air temperature increase of 3°C. A quadrupling of the concentration, which is a forcing twice as large as a doubling, is predicted to cause a temperature increase of 6°C. Accurate knowledge of climate sensitivity is considered to be so important that it has been referred to as the 'holy grail' of climate science (Pierrehumbert, 2013).

Climate scientists, however, have long realised that climate changes are not always proportional to the radiative forcing. A classic example is the phenomenon of deglaciation, which terminates ice ages. The latest deglaciation occurred between 20,000 and

15,000 years ago and it is believed that deglaciations involve a self-perpetuating feed-back loop. Once a critical amount of ice has melted, early in the process, any further ice loss causes even more ice loss until the global climate reaches interglacial conditions.

Until about the year 2004, climate scientists used the phrase 'critical phenomena' to describe dynamics associated with threshold effects which would lead to qualitative changes in the system state such as deglaciation (Rial, 2004). This concept of critical phenomenon is reasonably well defined in physics, where it has a meaning distinct from its usage in social sciences. In physics, a system becomes 'critical' if it is on the cusp of transforming to a new condition. 'Critical thresholds' identify system states or forcing levels which are about to trigger a critical phenomenon.

Since 2005, however, prominent climate scientists have used 'tipping point' to evoke the prospect of accelerated, or irreversible, global warming (see the historical review in Russill, 2015). The phrase, mainly used first in the mass media, has since permeated the mainstream academic literature, the language of politicians and funding agencies. A number of commentators have, however, worried that the notion of tipping point is ambiguous and politically loaded, with suspicious origins (Russill and Lavin, 2012), prone to convey fear and anxiety, and triggering maladaptive responses by the public and policymakers (Russill, 2015). While these criticisms need to be considered seriously, I will argue in this essay that the concept of tipping point is nevertheless helpful for describing and communicating possible climate futures. It bridges knowledge established with climate models with the concerns of climate governance and accountable risk man-agement. However, its use in science needs to be justified and adequately framed.

The concept of bifurcation and critical phenomena in mathematics and physics

Let me first explain why I believe that we need another concept, besides bifurcation and critical phenomena. The latter are well defined in mathematics and physics. In physics, the principle that a force causes an effect is classically encoded as an equation of the form $F = \dot{x}$. The left-hand side of this equation is the force (F), that is the cause of a variation in x. The dot above x means that the quantity equated to F is the *rate of variation* of the system state x. The force F causes a change, or difference, in x. For this reason, the equation $F = \dot{x}$ is called a *differential equation*. Physicists deal with such equations all the time. In general, they attempt to express F as a function of the state of the system and of various environmental factors that may be considered as external to the system. Let us call these environmental factors E. We therefore have $F(x, E) = \dot{x}$.

This equation defines what mathematicians call a *dynamical system*. Once the state of x is given at a specific time, the resolution of the equation allows one to determine the value of x at any other time, in the past or in the future. The theory of dynamical sys-tems is very rich and mathematicians have been able to formalise some not-quite intui-tive notions such as chaos, attractor and bifurcation (Strogatz, 2014, is a classical textbook). The notion of an *attractor*, for example, captures the idea that the state of the system, when left on its own in a given environment, evolves to a stationary state. The notion of *bifurcation* encodes the idea that in certain circumstances, a small change in the environment E may cause a qualitative change in the system's attractor.

The bamboo stick provides a natural example. It stays rigid when pressed along its vertical axis, but bends suddenly when the force being applied on the stick exceeds a given threshold. The bamboo stick is a reasonably simple physical object and physicists can provide a dynamical system model of it of the form $F(x, E) = \dot{x}$. This describes the movements of the bamboo stick fairly accurately. The sudden bending of the bamboo stick can then be identified as a 'bifurcation' in this mathematical model, and be associated with a critical level of the force exerted on it. Thus, with this concept of bifurcation, we have captured the idea that at some point a small change in the environment (here, the force applied on the bamboo) causes a qualitative change in the system's state (the stick breaks).

Climate physicists and meteorologists use dynamical systems to understand and predict climate change. The simplest models with just a few variables are designed to idealise a specific phenomenon, such as the ocean circulation or climate-vegetation interactions. For example, Brovkin et al. (1998) provided a simple model representing the interactions between vegetation and precipitations in the Sahara. This model presents a bifurcation point, separating a 'green' Sahara, with extensive vegetation cover, from a 'white' Sahara, similar to today. Brovkin and colleagues suggested that the desertification which occurred 6,000 years ago can be interpreted as the expression of this bifurcation.

For predicting future climate changes, however, climate scientists use models with millions of variables, which resolve the flows of the atmosphere and oceans. In such models, it is generally very hard to identify a bifurcation as mathematicians define it, but one *can* recognise behaviours that are akin to those observed when a bifurcation is approached or crossed in a *simple* dynamical system (Bathiany et al., 2016). These studies have led climate scientists to consider that *critical transitions* in regional climates and vegetation states may occur in the course of future climate change.

Why then would we need another descriptive term such as a tipping point? To understand this, let us first observe that the language and theory of dynamical systems have been used well beyond the domain of physics. In the early twentieth century, the mathematicians Alfred Lotka and Vito Volterra independently proposed a biological model capturing the dynamics of prey and predator, thereby opening the field of what we call today mathematical biology. Their dynamical system generates all sorts of interesting and not-so-intuitive mathematical phenomena, such as self-sustained oscillations and sensitive dependence on initial conditions. The formation of patterns of vegetation or zebra stripes have also been modelled and understood as the realisation of a bifurcation in a dynamical system. The dynamical system framework and language have turned out to be so powerful that it has become common to use them for explaining phenomena involving human interactions, such as urban congestion, opinion dynamics and racial segregation.

However, biological systems, and even more so social systems, are not bamboo sticks. They are multifaceted, complex systems which can be approached and modelled from many different points of view. Biological systems involve signal emission and recognition, learning and strategic adaptation. Humans use cultural symbols, hold diverse values and their behaviour is influenced by the social context within which they live their lives. Quite obviously, a dynamical representation of a living system (i.e. a model) can only capture a very narrow and restricted aspect of its causal structure. Indeed, the complexity of living systems (and, arguably, some social systems) is in part related to their ability to decode environmental signals and use them as proxies for the future. In other words, they decode information for *anticipating*. However, if the circumstances change in such a way that signals are not adequately interpreted, then the learning processes that have been accumulated under a previous set of conditions are at risk of causing inadequate responses to the new circumstances. This explains why, facing a new, sudden and unexpected situation, humans and other living creatures are at risk of responding in a way that is said to be 'maladaptive'; how they will react is hard to predict. This phenomenon cannot be fully captured with the dynamical systems framework, because it is not possible to encode in the system state 'x' all the learning history, all the context associated with signal recognition or all the anticipatory strategies of the human actors.

We therefore need a specific term, besides bifurcation or critical transition, which denotes any crisis likely to cause a failure of anticipation and inadequate responses. I will now suggest that the concept of tipping point can be used to this end.

The controversial origins of tipping points

It is difficult to trace the first use of tipping point in the English language and we may simply assume that it began its career as an informal metaphor. Weighing scales near their equilibrium point can be 'tipped' and we all feel situations where an extra nudge may qualitatively change a situation—as in the aphorism 'the straw that breaks the camel's back'. In the 1970s, the notion 'tip point' was used to depict mechanisms of racial segregation and spatial structuring in US cities, a phenomenon which was later modelled with dynamical systems and game theory. The topic of racial segregation and urban violence is at the heart of journalist Malcolm Gladwell's bestselling book *The Tipping Point* (Gladwell, 2000). He used this metaphor to argue that small environmental changes which can be implemented by policymakers (so-called Band-Aid solutions) may have important and beneficial effects on social structures.

In 2005, after the disaster caused by Hurricane Katrina striking New Orleans, climate scientist James Hansen stated that 'we are in the precipice of climate tipping points beyond which there is no redemption' (cited by Russill, 2015, p. 429). One year earlier, German Earth system scientist John Schellnhuber said in a BBC interview[1] that 'we should have a much better understanding of these tipping points' and he began to promote a 'tipping point map'. One version of this tipping point map, published in the prominent journal *Proceedings of the National Academy of Sciences*, has received thousands of academic citations (Lenton et al., 2008). We see here that the tipping point metaphor has been used to depict different notions—a critical decision, a point in time, or a critically sensitive system—but in all instances the political connotation of the concept is palpable. Invoking a tipping point implies something serious that demands urgent attention.

I propose to acknowledge the politically loaded origins of the tipping point, and build on these origins to establish it as a concept that constructively connects dynamical system language with the necessity of far-thinking climate governance.

A proposal for a definition of tipping points

Before proceeding further let me be clear about one thing. Climate scientists have *not* established a point in the foreseeable future where, at the planetary scale, global warming would be runaway—i.e. the decoupling of the magnitude of the forcing to the system's response. The prospect of a domino cascade of abrupt regional transitions putting the world's climate 'out of control' has been only hinted at (Steffen et al., 2018). It is an imaginary scenario that should be investigated seriously by scientists, but it has yet to be established either theoretically or empirically as a real possibility.

However, what climate scientists have established is the possibility of a sudden transition at least at the regional level (Bathiany et al., 2016). Such transitions have been legitimately considered as 'reasons for concern' because they would seriously impact ecological communities, human welfare and geopolitical dynamics (O'Neill et al., 2017). In parallel, we face the prospect that gradual changes in climate may cause sudden changes in social organisations that will cause population movements, and changes of habits, cultures and behaviour (see **Chapter 4**). Such reorganisations have sometimes been referred to as 'social tipping points' (Bentley et al., 2014). Rapid climate change and consequent disruptive social responses constitute moments during which normal modes of operations are seriously challenged. Such circumstances demand urgent action in a context of high uncertainty and where social values are at stake. These are times of potential crisis (see **Chapter 1**).

In such situations the tipping point can be a useful metaphor if the intuitive ideas that it conveys—suddenness, long-term commitment and challenging decision context—can be used constructively. In such a situation one can say that it is *generative* (Russill and Nyssa, 2009, citing D.A. Schön). I consider that this generative role can be fulfilled if the idea of tipping point is linked with the notion of accountable decision-making. We need to remember that whether we are farmers, bankers, parents and so on, we are all decision-makers. We all face decisions related to our changing environment and changing climate. Important decisions are and will also be taken at collective levels, in corporate industries, governments and non-governmental bodies (see **Chapter 12**), with potentially important and far-reaching consequences.

However, as climate changes (as too do other environmental factors), the risk of inadequate decisions increases for two reasons. The first one is quite obvious: climate change is partly unpredictable, especially if and when mechanisms of abrupt change are to be triggered—for example the evolution of large ecosystems is particularly hard to predict. The other reason is linked to the complexity of social systems. Which actions turn out to be valuable or rewarding are affected both by the climate change context and by the aggregated decisions taken at individual and collective levels. The decision-maker therefore needs to be able to recognise and anticipate circumstances that suddenly challenge her knowledge of what is a good or a bad decision. These circumstances can be thought of as 'tipping points' and the use of this metaphor helps to focus attention on them.

From the bifurcation to the tipping point

Three examples of increasing relevance will illustrate this proposal.

Anthropogenic emissions of greenhouse gases (GHGs) have now committed Earth's climate to a long interglacial period and have broken the oscillation of glacial-interglacial cycles. Indeed, CO_2 concentrations in excess of only 300–350 ppm would forbid any possibility of glacial inception within 50,000 years (Berger and Loutre, 2002); present concentration is about 420 ppm. This phenomenon of glacial-interglacial oscillations can be modelled and analysed in the language of non-autonomous dynamical system theory. Skipping the next glacial inception can be described, in this language, as a bifurcation. While such a far-future outcome admittedly has some philosophical resonance about, say, the status of humankind on Earth, its impact on decision schemes and values has so far been, let us say, nonexistent. Hence, under my reasoning, with regard to this outcome of postponing the next ice age we can speak of bifurcation or critical transition, but not of a tipping point.

With CO_2 concentrations above 400–500 ppm, the eventual melting of the Greenland ice sheet likely becomes inevitable and probably part of the West Antarctica ice sheet also (Edwards et al., 2019). The resulting sea-level increase will require decisions about infrastructure planning and involve the relocation of populations living in coastal areas. The socio-economic consequences are likely to affect people's habits, decision rules and value systems. Hopefully, the process will be slow enough to permit adequate adaptation of decision schemes and hence one might still avoid speaking of a tipping point.

My third example is inspired by a study by Defrance et al. (2017). They considered the so-called **RCP8.5 scenarios** in which CO_2 concentration exceeds 1000 ppm by the end of the century. Under this scenario, and based on a review of the literature, these authors considered as plausible the theory that the flow of meltwater coming from Greenland may be large enough to cause a substantial change in the North Atlantic ocean circulation. They then performed simulations with a climate model, and found that these circulation changes would aggravate the aridity in the African Sahel. They combined these simulations with demographic projections and concluded that climate change would in this case force the relocation of hundreds of millions of people. This is an example of a highly uncertain scenario that is nevertheless considered as a serious possibility. Quite obviously, the environmental and socio-economic development of such an event are highly unpredictable and sure to challenge most decision schemes. So I would argue that this outcome is clearly a tipping point that deserves anticipatory action and planning.

Conclusion

Adaptation to, and mitigation of, climate change not only requires incremental measures. It also requires thoughtful consideration of events that would suddenly trigger mass social disruption, challenge value systems and generate partly unpredictable outcomes. Tipping point is a useful concept to communicate and plan such possible futures if it is meant to alert decision-makers to the serious possibility of scenarios which could otherwise trigger inadequate responses. In my argument, the concept of tipping point is distinguished from the technical concept of bifurcation. The latter is a mathematical singularity which occurs in a certain class of models—dynamical system models—that are specified to capture the evolution of an idealised system. The concept of tipping point addresses the point of view of the decision-maker who needs to act with limited information in a complex and changing environment.

In my view, the notion of tipping point has also immediate implications for the promotion of scientific research programmes. Planning for possible climate futures requires more than expressing a likely range of climate sensitivity and temperatures. Similar to aeroplane passengers being told how to act in case of emergency, highly challenging scenarios need to be anticipated and have their early warning signs decoded. This implies that scientists should proactively investigate tipping point scenarios, even when they do not spontaneously emerge from standard experiments with state-of-the-art models. Scientists need to do so, not for the sake of telling fearful stories, but because proper anticipation of extreme scenarios is a genuine part of accountable decision-making.

NO: It misleads as to the nature of climate change (*James Annan*)

What are tipping points?

The concept of tipping points was introduced into climate science through the notion of abrupt climate change whereby gradual changes in radiative forcing were hypothesised to have the potential to trigger abrupt large-scale changes in the physical climate system. The paleoclimate record contains evidence of abrupt changes that may be (at least locally) several degrees Celsius in magnitude in as little as a decade (Broecker et al., 1985), although these have tended to be linked to climatic states with much larger land ice sheets than at present. This glacial background state is thought to be an important factor in facilitating these abrupt changes, for example through ice sheet binge-purge oscillations.

The mathematical principles of hysteresis and bifurcation theory (of which catastrophe theory is a special case) are well understood (e.g. Saunders, 1980) and models of the climate system have been shown to exhibit multiple stable states (Manabe and Stouffer, 1988). This combination of observational and modelling evidence indicates that abrupt changes between different climatic equilibria are *in principle* possible under a steady change in forcing. The terminology of tipping points was originally introduced in 2005 into the climate change discourse in this context (see review of Russill, 2015).

With respect to future climate change, the concept of tipping points has more recently been applied to regional climate subsystems which may exhibit some qualitative divergence over time due to a threshold effect or a sensitive dependence on the forcing trajectory (Lenton et al., 2008). This is a looser characterisation of tipping points. It implies that any time we define or identify a qualitative or quantitative threshold in the description of the climate (sub-)system under anthropogenic forcing, we necessarily obtain a 'tipping element'. This is because the set of anthropogenically forced trajectories of regional climate that do not exceed this threshold may be arbitrarily close to those that do. Thus, these tipping elements are simply components of the climate system where researchers consider some degree of change—for example brought about by greenhouse gas emissions—to eventually result in a qualitatively different climatic state.

There is no notion of rapidity in this definition—formally, Lenton et al. (2008) use a 1000-year time-scale to explore tipping points—so these changes are not necessarily abrupt on human time-scales. Indeed, it was noted by Lenton et al. (2008) that most of their proposed tipping elements have a response time-scale which is slow relative to the current forcing changes. For example, depending on the total anthropogenic emissions of CO_2 the climate system may well reach a state at which the Greenland ice sheet has melted in an essentially irreversible manner. But even if we have already exceeded the temperature threshold at which this would eventually happen (which experts consider unlikely, but perhaps possible), it is likely to take several centuries to occur. In human terms this cannot be said to be rapid.

In such a context, it is not so clear whether using the terminology of 'tipping' is aiding or hindering public understanding of climate science. If greenhouse gas forcing continues to rise indefinitely, the Greenland threshold will inevitably be exceeded. Conversely, if the rise in forcing could be halted and ultimately reversed prior to the completion of the melt, then we may save the Greenland ice sheet. The only way to predict the long-term trajectory of the system in this case would be to explicitly calculate it with a model which takes into account the speed of response of the climate system.

In this essay, I will argue that, while the long-term equilibrium response is undoubtedly of interest to climate scientists, use of the 'tipping' terminology in cases such as these does not seem helpful for informing the public about climate science or for developing policy responses to climate change.

Public understanding of tipping points

The tipping terminology and concept is inseparably linked to *rapidity of change* in general popular usage and public consciousness outside of the climate research sphere (Gladwell, 2000; Russill, 2015). Therefore it is perhaps not surprising that the public interpretation of tipping points tends towards the melodramatic. When Swedish schoolgirl Greta Thunberg gave a speech to MPs in the UK Parliament in April 2019, she said:

> Around the year 2030, 10 years, 252 days and 10 hours away from now, we will be in a position where we set off an irreversible chain reaction beyond human control, that will most likely lead to the end of our civilisation as we know it. That is unless in that time, permanent and unprecedented changes in all aspects of society have taken place, including a reduction of CO_2 emissions by at least 50%.[2]

One presumes that in these comments she was representing the scientific consensus appropriately as she understood it (aside from the obviously exaggerated precision of the time-scale) and indeed hundreds of climate scientists explicitly gave her their support. And in case anyone thinks that allowing 10 years to halve our emissions is insufficiently challenging, the Extinction Rebellion protest group have subsequently demanded that the UK reaches net zero emissions by 2025. I must emphasise that this is not just a matter of poor wording from one schoolgirl; the United Nations Secretary General António Guterres has also talked of 'runaway climate change'[3] if we do not change course immediately. These claims have very limited scientific support as we shall explore in the next section.

What is the scientific consensus on tipping points?

Scientists have explored the possibility of tipping points in future climate change simulations (Lenton et al., 2008; Steffen et al., 2018), but have not found much evidence in support. As the 2013 IPCC Report (Cubasch et al., 2013: 129) summarised: 'there is no evidence for global-scale tipping points in any of the most comprehensive models evaluated to date in studies of climate evolution in the twenty-first century.' More recently, the IPCC considered the question of 1.5°C of warming (IPCC, 2018) and made the point that immediate and rapid reduction in emissions would be required in order to avoid exceeding this level of warming. This appears to have been widely interpreted as giving us 12 years to save the planet. However, as one of the co-authors of this 2018 IPCC Report, Myles Allen, also said:

> So please stop saying something globally bad is going to happen in 2030. Bad stuff is already happening and every half a degree of warming matters, but the IPCC does not draw a "planetary boundary" at 1.5°C beyond which lie climate dragons.[4]

Climate change is primarily an incremental and long-term process. Every year of postponement in serious reduction of emissions will lead to an increase of around 0.02–0.03°C in the maximum global temperature rise, with this extra annual commitment itself rising gradually over time. Most of the warming effect from GHG emissions actually occurs fairly rapidly, reaching its maximum in about a decade after their release into the atmosphere (Ricke and Caldeira, 2014)—although of course the planning and construction of energy infrastructure also often implies a commitment to future emissions extending over a significantly longer time period. The prospect of exceeding 3 or 4°C of warming in the coming century if emissions are unchecked should certainly be sufficiently worrying to motivate policies to avoid this. Talking of commitment in the context of slow future changes implicitly assumes that radiative forcing of the climate system cannot be significantly reduced below current levels. However, if we can stabilise atmospheric concentrations—implying a massive reduction in global emissions by 80–90% from current levels —we could presumably also go a little beyond mere stabilisation of these concentrations and actually reduce them. Even net zero emissions would result in a gradual reduction in atmospheric CO_2, albeit the concentration would remain elevated significantly above pre-industrial levels for many thousands of years.

It would of course be unwise to simply assume that technology will solve our problems at some unspecified point in the future. But in talking of commitment to future changes over long time-scales (for example, several centuries of sea-level rise) it must be recognised that these commitments are not absolute, but instead conditional on future *in*action. We can define any thresholds we choose as denoting 'danger', but the basic processes of climate change are much closer to being continuous—i.e. incremental—than discontinuous. The dominant features of the problem of climate change are the long time-scale and differential regional impacts which create serious issues of intergenerational and international inequities (see **Chapter 11**), rather than unpredictable thresholds, abruptness and irreversibility of change. The concept of tipping point is therefore not a helpful one.

Why is public understanding so poor?

As noted above, the use of tipping point differs crucially in climate science from its usage in the wider public sphere. It is hardly surprising that members of the public will not always understand the specific way the tipping point concept is used in climate change research; they may also have difficulty in grasping time-scales of change which are multi-centennial in nature (we should note in passing that projecting backwards 300 years also takes us to a society and economy vastly different to our current situation). It also seems that some scientists have played up this misunderstanding, by exaggerating the impact and immediacy of their storylines. For example, the 'hothouse Earth' hypothesis of Steffen et al. (2018) became the climatic catastrophe of Johan Rockström:

> We are the ones in control right now, but once we go past 2 degrees, we see that the Earth system tips over from being a friend to a foe. We totally hand over our fate to an Earth system that starts rolling out of equilibrium.[5]

Similarly, the ambitious emissions reduction hypothesised by the IPCC in their 2018 report on 1.5°C was widely reported as implying we had a mere 12 years to save the climate. Let there be no doubt whatsoever, the massive preponderance of scientific evidence is on the side of Myles Allen rather than Johan Rockström (see the above quotes).

However, with the message of the IPCC over several decades being seemingly insufficiently powerful to generate the desired political response, it seems that some climate activists have weaponised the language of tipping points in the hope that this will alarm and motivate the public (Asayama et al., 2019).

Discussion

In reality there is not a binary outcome where an empirical threshold demarcates between 'safe' and 'dangerous' climates. Encouraging this misunderstanding is misleading and potentially damaging. Back in 2008, some campaigners were telling us we had 100 months to save the planet (the countdown website www.onehundredmonths.org set up at that time is no longer actively updated).[6] There may be a limit to the number of times we can 'cry wolf'. When we get to 2025 and emissions have not been eliminated, anyone who believed that this was some firm deadline beyond which lay disaster may feel somewhat demotivated.

In fact the climatic difference between immediate stringent mitigation of emissions starting now, and equally stringent mitigation starting in 2030, is roughly a quarter of a degree of warming in the long-term. This point is emphatically not intended as an argument in support of such a delay, merely a plea that we address the costs and benefits of policy decisions in an honest manner. A quarter of a degree of additional warming will bring a range of harms and damages to ecosystems, as well as costs to the human economy that can be accounted for in financial terms (see **Chapter 5**). A further decade of delay beyond that, with its further concomitant quarter degree or so of warming, would lead to somewhat greater additional harm and costs, and so on. The prospect of climate change that (if unchecked) could exceed 3 or 4°C in the coming century is certainly something many people would reasonably be worried by. But there is no scientific reason to believe there is a particular tipping point or threshold demarcating a safe space with low harm, from a catastrophic collapse of civilisation. While there is legitimate scientific interest in exploring the potential of the climate system to behave in abrupt and/or irreversible ways, the relevance of the tipping point concept to our current predicament is tenuous at best. Language of emergency and extinction crowds out more realistic discussion of the problem and our options in dealing with it.

The concept of the **Overton Window** (Lehman, 2014) has recently attracted attention as a metaphor for understanding the range of policies that are considered viable in political discourse. According to this theory, promoting a position on climate change that seems currently extreme (i.e. outside the window of acceptability) may help to shift mainstream opinion in the preferred direction. Some may therefore attempt to excuse the weaponising of the language and terminology of tipping points through the hope that this may shift the Overton Window to a position more favourable to mitigation policies. But an alternative outcome of emphasising extreme positions is also possible. It may be that the Window is 'broken' to the extent that there is strong polarisation between alarmism and climate change denial, with little opportunity for honest and constructive debate over climate change in the space between these two extremes. When faced with a difficult or impossible task, fear and paralysis, or even increased denial, may result (Brügger et al., 2015; Haltinner and Sarathchandra, 2018; Wolfe and Tubi, 2019). Rather than making ever-more-extreme claims about urgency and impending catastrophe, I would argue that the climate change problem should be more honestly described as one of gradual, but significant, commitment to ever larger changes, which will certainly cause future harm in many situations.

Conclusions

There is legitimate scientific interest in understanding the potential of the climate system to behave in complex ways including abrupt changes and thresholds. However, there is limited evidence that this has much relevance to our future on policy-relevant time-scales. 'Tipping' terminology is also attractive for some forms of advocacy through its attention-grabbing nature in public discourse. Despite this, or perhaps because of it, we should be wary of invoking the term in situations where it does not genuinely apply.

The late climate scientist Steven Schneider talked of the balance between being effective and being honest and the hope that scientists would manage to achieve both of these things.[7] Focusing on thresholds and deadlines for rhetorical purposes, using the concept of tipping point, risks failure on both counts. Exaggerating the immediacy and severity of the problem may be intended to motivate action, but it is not clear that this will be effective. Our interests would be better served by making serious efforts to accelerate reductions in emissions, rather than in hyping up fear over ever-more-extreme scenarios of climate catastrophe. A primary responsibility of scientists is to be honest and avoid misleading the public in our communication. The creation of artificial deadlines and threats of imminent doom, conjured up by tipping points, fails on both fronts.

Further reading

Kopp, R.E., Shwom, R.L., Wagner, G. and Yuan, J. (2016) Tipping elements and climate-economic shocks: pathways toward integrated assessment. *Earth's Future*. **4**: 346–372.

This article suggests a terminology that distinguishes between climate tipping points, climatically sensitive social tipping elements, and climate-economic shocks, and how they may be incorporated into climate change risk analysis. The authors also provide further references to the possibility of maladaptive responses to large or sudden climatic changes.

Russill, C. (2015) Climate change tipping points: origins, precursors, and debates. *WIREs: Climate Change*. **6**(4): 427–434.

This article offers a useful review of the origins, precursors and main advocates of climate change tipping points and the debates that the tipping point concept has occasioned. Russill advocates for a deeper understanding of dynamical systems theory and its origins—both mathematical and metaphorical—for addressing the value of the tipping point concept in policy discourse.

Scheffer, M. (2009) *Critical Transitions in Nature and Society*. Princeton, NJ: Princeton University Press.

This book offers a widely accessible introduction to the idea of 'critical transitions', showing how the notions of bifurcation and critical phenomena are used to explain and predict rapid or irreversible changes outside the domain of physics.

The 'TIPES' project—Tipping Points in the Earth System (https://tipes.sites.ku.dk/).

This website is an example of a collaborative research action funded by the European Union. It gathers together mathematicians, physicists and climate scientists and promises to 'communicate to policymakers in a manner that facilitates decisions and their implementation'.

Follow-up questions for use in student classes

1. Based on the above discussion, would you consider that there is one global, planetary tipping point?

2. What is the value, and the limit, of the concept of bifurcation to describe abrupt climate change? In what sense does the concept of tipping point address these limitations?
3. How do you address the criticism that talking about climate tipping point serves a political agenda and builds anxiety and fear?
4. What time-scale of change is 'rapid' for you? How do you think your time-scale of rapid change compares with that imagined by scientists, politicians or the media?

Notes

1 See: Kirby, A. Earth warned on 'tipping points.' *BBC News* 2004, August 26. Available at: http://news.bbc.co.uk/2/hi/science/nature/3597584.stm Accessed 16 July 2019.
2 See: Thunberg speech to MPs: www.theguardian.com/environment/2019/apr/23/greta-thunberg-full-speech-to-mps-you-did-not-act-in-time Accessed 16 July 2019.
3 See: www.scmp.com/news/world/europe/article/2183670/we-are-losing-race-climate-change-un-chief-antonio-guterres-warns Accessed 16 July 2019.
4 See: https://theconversation.com/why-protesters-should-be-wary-of-12-years-to-climate-breakdown-rhetoric-115489 Accessed 16 July 2019.
5 See Rockström comments to *BBC News:* www.bbc.co.uk/news/science-environment-45084144 Accessed 16 July 2019.
6 See: 'Deadline-ism': when is it too late? Available at: https://mikehulme.org/deadline-ism-when-is-it-too-late/ Accessed 16 July 2019.
7 See: 'Don't bet all environmental changes will be beneficial', APS News, 5, 1996. Available at: www.aps.org/publications/apsnews/199608/environmental.cfm Accessed 16 July 2019.

References

Asayama, S., Bellamy, R., Geden, O., Pearce, W. and Hulme, M. (2019) Why setting a climate deadline is dangerous. *Nature Climate Change.* **9**(8): 570–572.

Bathiany, S., Dijkstra, H., Crucifix, M., Dakos, V., Brovkin, V., Williamson, M., Lenton, T. and Scheffer, M. (2016) Beyond bifurcation: using complex models to understand and predict abrupt climate change. *Dynamics and Statistics of the Climate System.* **1**(1):dzw004, doi: 10.1093/climsys/dzw004.

Bentley, R.A., Maddison, E.J., Ranner, P.H., Bissell, J., Caiado, C.C.S., Bhatanacharoen, P. and Clark, T. (2014) Social tipping points and Earth Systems dynamics. *Frontiers in Environmental Science.* **2**(35), doi:10.3389/fenvs.2014.00035.

Berger, A. and Loutre, M.F. (2002) An exceptionally long interglacial ahead? *Science.* **297**: 1287–1288.

Broecker, W.S., Peteet, D.M. and Rind, D. (1985) Does the ocean-atmosphere system have more than one stable mode of operation? *Nature.* **315**: 21–26.

Brovkin, V., Claussen, M., Petoukhov, V. and Ganopolski, A. (1998) On the stability of the atmosphere-vegetation system in the Sahara/Sahel region. *Journal of Geophysical Research.* **103**: 31613–31624.

Brügger, A., Dessai, S., Devine-Wright, P., Morton, T.A. and Pidgeon, N.F. (2015) Psychological responses to the proximity of climate change. *Nature Climate Change.* **5**(12): 1031–1037.

Cubasch, U., Wuebbles, D., Chen, D., Facchini, M., Frame, D., Mahowald, N. and Winther, J.-G. (2013) Introduction. In Stocker, T., Qin, D., Plattner, G.-K., Tignor, M., Allen, S., Boschung, J., Nauels, A., Xia, Y., Bex, V. and Midgley, P., eds. *Climate Change 2013: The Physical Science Basis. Contribution of Working Group I to the Fifth Assessment Report of the IPCC.* Cambridge: Cambridge University Press. pp.119–158.

Defrance, D., Ramstein, G., Charbit, S., Vrac, M., Famien, A.M., Sultan, B. and Swingedouw, D. (2017) Consequences of rapid ice sheet melting on the Sahelian population vulnerability. *Proceedings of the National Academy of Sciences.* **114**(25): 6533–6538.

Edwards, T.L., Brandon, M.A., Durand, G., Edwards, N.R., Golledge, N.R., Holden, P.B., Nias, I.J., Payne, A.J., Ritz, C. and Wernecke, A. (2019) Revisiting Antarctic ice loss due to marine ice-cliff instability. *Nature.* **566**(7742): 58–64.

Gladwell, M. (2000) *The Tipping Point: How Little Things Can Make a Big Difference.* New York: A Back Bay Book: Little Brown.

Haltinner, K. and Sarathchandra, D. (2018) Climate change skepticism as a psychological coping strategy. *Sociology Compass.* **12**: e12 586.

IPCC (2018) Summary for Policymakers. In: *Global Warming of 1.5°C. An IPCC Special Report on the impacts of global warming of 1.5°C above pre-industrial levels and related global greenhouse gas emission pathways, in the context of strengthening the global response to the threat of climate change, sustainable development, and efforts to eradicate poverty* [Masson-Delmotte, V., P. Zhai, H.-O. Pörtner, D. Roberts, J. Skea, P.R. Shukla, A. Pirani, W. Moufouma-Okia, C. Péan, R. Pidcock, S. Connors, J.B.R. Matthews, Y. Chen, X. Zhou, M.I. Gomis, E. Lonnoy, T. Maycock, M. Tignor and T. Waterfield (eds.)]. World Meteorological Organization, Geneva, Switzerland, 32 pp.

Lehman, J. (2014) A brief explanation of the Overton Window. Available at: www.mackinac.org/overtonwindow#Explanation.

Lenton, T.M., Held, H., Kriegler, E., Hall, J.W., Lucht, W., Rahmstorf, S. and Schellnhuber, H.J. (2008) Tipping elements in the Earth's climate system. *Proceedings of the National Academy of Sciences.* **105**: 1786–1793.

Manabe, S. and Stouffer, R.J. (1988) Two stable equilibria of a coupled ocean-atmosphere model. *Journal of Climate.* **1**: 841–866.

O'Neill, B.C., Oppenheimer, M., Warren, R., Hallegatte, S., Kopp, R.E., Pörtner, H.O. and Scholes, R. (2017) IPCC reasons for concern regarding climate change risks. *Nature Climate Change.* **7**(1): 28–37.

Pierrehumbert, R.T. (2013) Hot climates, high sensitivity. *Proceedings of the National Academy of Sciences.* **110**: 14118–14119.

Rial, J. (2004) Abrupt climate change: chaos and order at orbital and millennial scales. *Global and Planetary Change.* **41**(2): 95–109.

Ricke, K.L. and Caldeira, K. (2014) Maximum warming occurs about one decade after a carbon dioxide emission. *Environmental Research Letters.* **9**: 124 002.

Russill, C. (2015) Climate change tipping points: origins, precursors, and debates. *WIREs: Climate Change.* **6**(4): 427–434.

Russill, C. and Lavin, C. (2012) 'Tipping point' discourse in dangerous times. *Canadian Review of American Studies.* **42**(2): 142–163.

Russill, C. and Nyssa, Z. (2009) The 'tipping point' trend in climate change communication. *Global Environmental Change.* **19**(3): 336–344.

Saunders, P.T. (1980) *An Introduction to Catastrophe Theory.* Cambridge: Cambridge University Press.

Steffen, W., Rockström, J., Richardson, K., Lenton, T.M., Folke, C., Liverman, D., Summerhayes, C.P., Barnosky, A.D., Cornell, S.E., Crucifix, M., Donges, J.F., Fetzer, I., Lade, S.J., Scheffer, M., Winkelmann, R. and Schellnhuber, H-J. (2018) Trajectories of the Earth System in the Anthropocene. *Proceedings of the National Academy of Sciences.* **115**(33): 8252–8259.

Strogatz, S. (2014) *Nonlinear Dynamics and Chaos: With Applications to Physics, Biology, Chemistry, and Engineering.* Boulder, CO: Westview Press Books.

Wolfe, S.E. and Tubi, A. (2019) Terror Management Theory and mortality awareness: a missing link in climate response studies? *WIREs: Climate Change.* **10**(2): e566.

3 Should individual extreme weather events be attributed to human agency?

Friederike E.L. Otto and Greg Lusk

Summary of the debate

This debate introduces students to the new science of weather event attribution and investigates whether the objective of declaring specific weather extremes to be either of human cause or not to be a desirable one. **Friederike E.L. Otto** shows how far weather attribution science has developed in recent years and argues that such studies are the only means to understand accurately what the impacts of human-induced climate change are today. They allow us to understand the relative importance of different meteorological and societal drivers of disasters. **Greg Lusk** emphasises some of the ambiguities and confusions in weather attribution claims and finds that the claimed social benefits of weather event attribution do not pass scrutiny. He argues that the attribution of specific individual events should be avoided in favour of analysing and communicating changes in *types* of extreme weather events.

YES: Attribution provides a realistic view of the impacts of climate change and can improve local decision-making and planning (*Friederike E.L. Otto*)

Introduction

In scientific reports and political debates, and to a large degree also in the media, the measure of global climate change around the world is global mean surface air temperature. This is the metric most frequently used to determine by how much humans are changing the climate by burning fossil fuels. Most prominently, the change in global mean temperature has been used as the quantity to measure and discuss a changing climate in the United Nations Framework Convention on Climate Change (UNFCCC) negotiations. Recently, the 2015 **Paris Agreement** on Climate Change states that limiting the increase in global mean temperature to well below 2°C above the pre-industrial level is the central goal of international climate policy and politics. It is, however, not the abstract measure of global mean temperature that will cause loss and damage from climate change. Rather it is the local and regional impacts of climate change that are primarily manifest through rising sea-level and the changing likelihood and intensity of extreme weather events.

Until recently it has not been possible to make the crucial link between anthropogenic global climate change and specific weather and climate-related high-impact events with scientific confidence. In recent years this has changed (Herring et al., 2018). Quantifying and establishing the link between global warming and individual weather events that often lead to significant social and ecological damage has been the focus of the emerging science of extreme event attribution. As with future climate projections, scientific knowledge is limited by the ability of climate models to reliably simulate weather events. Yet an increasing number of extreme events can now be robustly attributed to human influence—in particular heatwaves and extreme rainfall events, but also some droughts and storms. Not unsurprisingly, however, when facing new scientific developments not everyone agrees that just because we *can* now attribute extreme weather events to human actions we *should* in fact do so.

While in some contexts such reservations might be appropriate, there are two compelling arguments in favour of attribution of individual weather events. First, even if a comprehensive inventory of the impacts of climate change today is impossible, extreme weather attribution helps us to understand better what human-induced climate change means at local and regional scales, today, for every one of us. Extreme weather attribution turns climate change from an abstract threat in the future to being a felt reality today. Second, and more importantly, is the ability of this science to disentangle the drivers of a damaging extreme event: human-caused climate change, natural climate variability or changes in vulnerability and exposure to extreme weather. Disentangling such factors allows a better understanding of the sources of risk and damage and, in turn, helps us determine how such damage can be reduced in the future. Scientific evidence of the relative importance of different drivers of meteorological disasters is essential to avoid playing uninformed blame games. Instead, it allows for well-informed debate about addressing and reducing risk.

In addition to these two arguments, it is also important to highlight the fact that from a scientific point of view much is learned when many scientists are trying to do something new. Improvements in scientific method and practice are made when many different scientists are working independently on the same problem and scrutinise each other's work.

Attributing extreme weather events

The term 'attribution' in the academic literature on climate change is not as clearly defined as it might seem. It was originally used to refer to the detection and attribution of observed long-term trends in climate to human actions. This application of attribution science used regression methodologies known as optimal fingerprinting that were first developed by Hasselmann (1997). Following common methodologies and best-practice protocols, the term attribution has been used more recently to explain an observed trend in a climate variable (Lehner et al., 2018), an impact indicator (Hansen et al., 2016) or an extreme weather event (National Academies of Sciences, 2016; Shepherd, 2016). It can also be used to attribute damages and impacts to changes in extreme weather events due to climate change.

Given the emerging nature of attribution science and the fact that only very few studies currently exist (in particular for the last category of attributing damages), 'best practice' guidelines have not yet been agreed upon in the scientific community. In the past this has led to apparent contradictions in the results of weather attribution studies. Most famously, perhaps, was the study of the heatwave over Russia during the summer of 2010. One study claimed this heatwave to have mainly natural origins, while a second one published shortly after stated that anthropogenic climate change had increased the likelihood of a record-breaking heatwave by a factor of five. This apparent contradiction was shown to result from the two studies asking different questions rather than from incompatible methodologies. The two studies were in fact complementary (Otto et al., 2012).

These early studies in the short history of the attribution of extreme weather events highlighted the importance of how exactly the question that is being answered in a study is framed. This question framing, together with the precise definition of the extreme weather event being attributed, determines the quantitative, and sometimes also the qualitative, outcome of the attribution analysis. New attribution studies began to be conducted using, developing and refining a range of methodologies (e.g. Herring et al., 2018). These methods continue to evolve. The fact that no single methodology exists and that it is hard to agree upon a single standard or best practice is frequently used as an argument *against* pursuing extreme event attribution. In particular the dangers of prematurely communicating the results of a near real-time attribution study are sometimes highlighted. However, I would argue that exploring the same, or almost the same, question using very different methodologies, models and datasets is not unique to extreme event attribution. It is commonplace in climate science. It is no different to reporting projections of future climate conditions simulated in different ways using a range of models, scenarios and model ensembles. In both contexts, the precise results depend on the models being used and the framing of the question, but this should not invalidate the attempt.

One could argue that in one sense weather attribution is different to future climate projection. In the case of extreme event attribution there is in principle a 'correct' answer—either the event was caused by human actions or it was not. In the case of future projections of climate, we know that none of them can be exactly 'correct'—the future will not follow exactly any of the human development scenarios that drive these future climate simulations. While this is undeniably true, it does not change the fact that whenever scientists communicate results of a scientific study, the final numbers are only meaningful in the context of the assumptions made and the exact question being asked. This is true whether it is an extreme event attribution study of the recent drought in the Western Cape of South Africa, the projection of future rainfall in East Africa in a world 2°C warmer than at the beginning of the industrial revolution or an assessment of crop yields in southern China over the next 20 years.

Important as this realisation is, it should not lead to abandoning quantitative climate science. Rather, it leads to improved communication of the assumptions, values and contexts that shape any scientific study. Arguably, the attribution of extreme weather events has helped the development of improved climate science communication rather than hindered it. Studying the features and causes of the same weather event in many different ways has made climate scientists think, and communicate, more carefully about the framing and definition of their quantitative studies of a complex chaotic system (Otto et al., 2015; Shepherd, 2016).

Probabilistic event attribution

Just because there are different ways of asking the attribution question does not mean that there cannot be an agreed-upon way of how to answer the question under a particular framing. The most frequently used method of attributing an extreme weather event to human-caused climate change is *probabilistic event attribution*. In this method, a type of extreme event that causes some significant local impact, defined as what has been observed, is analysed with respect to its occurrence frequency and intensity. Given that extreme weather events are by definition rare, it is not possible to determine the likelihood of an event occurring from observations alone. The historic record will contain too few of them to establish robust likelihoods. For this reason, probabilistic event attribution studies employ statistical models and state-of-the-art climate models to simulate possible weather in two hypothetical worlds: a world like the real world (i.e. with human influences on climate) and a world in which there are no human influences on climate. While we do not have observations of this latter world, we do know very well how much carbon dioxide and other greenhouse gases have been emitted since the beginning of the Industrial Revolution. This enables us to use computer models of climate to simulate possible weather events in the world that *might have been* (i.e. the world without human-induced climate change) in the same way as we simulate possible weather events today.

The likelihood of the weather event of interest is subsequently estimated in these two worlds. In some cases, this likelihood can be extremely rare. For example, in the case of the rainfall associated with Hurricane Harvey in 2017, this was on the order of a 1-in-10,000 year event (Risser and Wehner, 2017). In other cases it might be a more common event, for example a drought in Kenya with a likelihood of 1-in-10 years or less (Uhe et al., 2017). Having determined the likelihood of the selected event in the simulated factual world (i.e. including all anthropogenic drivers), the likelihood of the similarly defined event is determined from the simulation of the counterfactual world *without* human-induced climate change.

Comparing the two likelihoods of the selected event in these two worlds then allows us to quantify the role of climate change in 'causing' the event. For example, a rainstorm like Storm Desmond hitting the north of the UK in 2015 is expected to occur once on average every 70 years. But in a world *without* human-caused climate change it would have been a rarer event, occurring on average once every 100 years. While still being roughly a once-in-a-lifetime event, attribution science allows us to say that the burning of fossil fuels has made the event about 40% more likely—i.e. a factor of (100–70)/70. Similarly, if we find from model simulations that a 1-in-10 year heatwave in northern Europe in the world without climate change would have a maximum temperature of 38°C, but the 1-in-10 year heatwave in today's climate would be 40°C, we would conclude that human greenhouse gas emissions have contributed 2°C to the magnitude of the heatwave.

There are some variations in exactly how this methodology is implemented, for example using coupled models or atmosphere-only models, using only statistical modelling based on observations or using a combination of all these approaches (Uhe et al., 2017). Nevertheless, scientists find that results are more robust to methodological and framing differences than they are to differing definitions of the event (Risser and Wehner, 2017). They also find that confidence in their findings is generally improved when methods and models are combined. Best practice in the field is emerging through systematic comparisons of extreme event attribution studies, in part driven by national weather services which seek to implement extreme event attribution within their operational service protocols.

Attribution using storylines

While the increase in attribution research has led to improved and more consolidated methodologies of probabilistic event attribution, this is not the only way attribution studies can be undertaken. Recently a different method of explaining the role of anthropogenic climate change in extreme weather event has been employed (Shepherd, 2016). Termed a 'storyline approach'—or *conditional* attribution—this asks a different question than 'what is the overall likelihood of a specific event occurring under different levels of greenhouse gas forcing?' Rather it isolates a particular extreme event that has occurred—say a hurricane—and then investigates by how much the general warming of the atmosphere has increased its magnitude. The numbers from such a conditional attribution study will differ from those of a probabilistic approach. One could argue that this can be very confusing for public audiences to understand, as happened for example with the 2010 Russian heatwave example cited earlier. It has been shown, however, that these two different methods of attributing weather events to human causation result in differences merely of degree, not kind (Jézéquel et al., 2018). Successful attempts to combine the storyline method with probabilistic framings of attribution have also been made (e.g. Cheng et al., 2018).

Much more important than the exact numbers that emerge from event attribution studies is that methodological diversity has led to a greatly improved understanding of the drivers of the extreme weather studied. By conducting attribution studies scientists are now able to say more confidently which of the projected changes in local and regional weather associated with anthropogenic climate change are already observable. Event attribution science is able to discriminate between which changes in weather have turned out to be very much as expected (e.g. changes in mid-latitude rainstorms and heatwaves), which are stronger than expected (e.g. rainfall associated with Atlantic hurricanes) and where projections and observations do not match (e.g. African rainfall extremes). Attribution studies thus highlight the local 'meteorological hot spots' of anthropogenic climate change, as well as highlighting areas where our scientific understanding is currently poor.

When climate change is not to blame

Crucially such studies highlight when human-caused climate change is *not* a major driver behind an extreme weather event and its associated damages. A good example of this is a drought that occurred in Brazil in the area around São Paolo in 2014. A prolonged lack of rainfall led to large impacts in the greater metropolitan area, in particular for the poorer parts of the population with outbreaks of cholera and other waterborne diseases. By conducting an attribution study (Otto et al., 2015), it was found that although evaporation increased slightly due to human-caused climate change, so too did rainfall. These two effects largely cancelled each other out, leading to no change in the availability of water—and thus drought risk—that could be attributed to human agency. The study furthermore revealed that from a meteorological point of view the drought event was not very extreme, occurring on average once per decade. Similar events had happened earlier in the twenty-first century and also in the 1970s. In both past cases, however, the impacts were much less severe. When investigating what else had changed, the authors found that not only had the population of the area increased greatly in the ten years since the previous drought, but water usage had increased almost exponentially. It was thus changes in human vulnerability and exposure that had turned a meteorological hazard into a catastrophic event, not human-caused climate change.

Another, much more complex example of a case where attribution studies have discovered that climate change is not to blame, at least not in a straightforward way, is from East Africa. Several attribution studies of recent droughts in the region have noted a decrease in rainfall in some areas, whereas other regions show no trend. In general, East Africa experiences large year-to-year natural variability in rainfall. In synthesising results from multiple studies, and across different assessment methods, it becomes clear that human-caused climate change is currently not a big driver of the recent lack of rainfall (e.g. Uhe et al., 2017). At the same time, increasing temperatures in the region *can* be attributed to human-induced climate change and this is often used to blame climate change as responsible for the droughts. But the effect of higher temperatures on drought occurrence in these regions is far from straightforward. And while water does indeed evaporate faster with higher temperatures, higher temperatures have little effect once all the water is gone.

The conclusion from these studies is that how human-induced climate change is affecting drought in East Africa strongly depends on the region and the season. A simple statement such as 'climate change is to blame' is not scientifically warranted. In the case of droughts in East Africa we have no evidence that climate change made these events more likely. The anthropogenic climate change signal in these weather extremes—while not completely absent—is likely to be small. The inconvenient, but honest, message from attribution science is that droughts are a natural part of this region's climate and will happen again in the near future for reasons unrelated to human-caused climate change. This message matters. In cases of high vulnerability a relatively moderate lack of rainfall can have a dramatic impact on people's lives, especially when political conflicts have left the country unequipped to provide assistance in times of drought (see **Chapter 4**).

Attribution studies conducted very shortly after an event has happened, and when decision-makers can rethink their disaster management strategy, can thus inform public awareness, policy and practice. Disentangling societal, meteorological and climatological drivers of disasters in a timely way creates a window of opportunity in which well-informed debate about addressing climate-related risks is possible, instead of playing an uninformed blame game. While vulnerabilities to climatic hazards are greatest in developing countries, these challenges are in no way limited to those parts of the world.

Conclusion

Attribution studies have *a priori* four possible outcomes: (1) an event can have been made *more* likely by human actions; (2) it can be made *less* likely; (3) the occurrence frequency can be *unchanged* due to anthropogenic climate change; or (4) because of insufficient data and tools it can be impossible to conduct an attribution study. The first outcome is often the one that receives most attention. In a world where climate change is still not high on the priority list of those in power, this is one of the two reasons to undertake attribution studies this essay has highlighted. Attribution studies are the only way to understand and demonstrate the impacts of specifically human-induced weather extremes today. These quantitative estimates not only help to identify the hotspots of climate change, but also the cost. For example, New Zealand's Treasury calculated on the basis of several attribution studies, using the most conservative estimate possible, that climate change has cost the country at least NZ$520 million in the last ten years.

It is the events falling in the other categories of possible outcomes, however, that provide the second, and are arguably the more important, reason to conduct attribution studies. It is extreme weather events where human-induced climate change is found *not* to blame where the emerging science is most powerful. This is because such results highlight where observed trends in extreme weather-related disasters—like the drying in Eastern Africa or high-impact events like droughts in the São Paolo region—have causes other than human influences on climate. These observed trends or events are therefore *not* harbingers of the future evolving along with climate change, but rather are risks that need to be addressed—and importantly *can* be addressed—independently of curbing greenhouse gas emissions. While on the one hand this highlights the *responsibilities* of local decision-makers, it also *empowers* local politicians and decision-makers. It shows that it is certainly not the case that anthropogenic climate change renders societal development gains in all parts of the world futile, nor that development is a lesser priority than mitigating climate change (see **Chapter 1**).

NO: Attributing individual extreme events to anthropogenic factors is not as useful as you might think (*Greg Lusk*)

Introduction

There is widespread scientific agreement that the 'warming of the climate system is unequivocal' and that the 'human influence on the climate system is clear' (Stocker et al., 2013: 4, 15). This human influence has been the primary driver of a rising trend in global mean temperature. Despite a broad scientific consensus on the reality of anthropogenic global warming (see **Chapter 9**), garnering sustained action to mitigate this warming has been difficult. According to the IPCC, past human action has likely already committed the climate system to at least a 1.5°C rise in global mean temperature above the pre-industrial level. Business-as-usual will produce even greater warming.

These measures are, however, abstract: no one ever feels global mean surface air temperature—it is a quantity constructed from specific surface temperatures people experience around the world. But people *will* feel the repercussions of a changing climate. We are already observing these repercussions: sea-level rise is already endangering small island nations; the loss of sea ice is shrinking the hunting grounds of northern indigenous peoples, threatening food security; and coastal communities face relocation due to permafrost thaw and erosion that makes their infrastructure unstable.

None of these specific realities of climate change is related to an extreme weather event. But extreme weather events are likely to cause their own harms and make such problems much worse. Sea-level rise is a perfect example. When a storm hits near a coastal area, a storm surge may occur, which is an abnormal rise of water much greater than the predictable high tides. If the sea level is higher than it otherwise would be due to a warmer planet, then the surge will be higher and stronger and more easily overcome the defences that protect civilisation (if the community has any such defences at all) causing greater damage. It seems obvious that we should be concerned with, and further investigate, the changing nature of extreme events in an altered climate.

Although investigating extreme weather events is important for understanding and fostering political action on climate change, I will argue here that attributing the occurrence of specific events to climate change is not. To support this conclusion, I will survey three reasons that have been cited for attributing specific events to climate change: (1) adaptive decision-making; (2) appropriating responsibility for damage; and (3) making climate change visible. I demonstrate that using the attribution of specific extreme events for these purposes is often more challenging than it would first appear. Furthermore, I show that there are sometimes different ways of achieving the same goals—equally well, if not better—which do not require the attribution of a specific extreme event. In short, we can use our knowledge of extreme events in general to get everything we want, without ever having to do the hard work of attributing the occurrence of specific weather events to human actions.

Attribution of specific individual extreme events

Heatwaves, droughts, deluges and hurricanes are all examples of extreme weather. Extreme weather event attribution attempts to quantify the extent to which human actions—typically the release of greenhouse gases and aerosols, as well as land use changes—have influenced extreme weather events. This 'influence' is often defined as a change in the frequency of occurrence. In other words, scientists try to answer the question: 'How has the likelihood of an extreme event occurring in some particular geographical area changed due to human actions?'

Answering this question scientifically often involves comparing the actual Earth (or some region of it) where such human actions are present, to an Earth where the actions were absent. Since scientists have no Earth free from human actions to observe, they often use computer simulation models of the climate where the human actions are easily adjustable. They compare simulations of a 'natural world' free from human actions to the simulation of an 'actual world' *with* human actions to assess how the occurrence of extreme events changes between the two worlds. If the probability of extreme weather occurrence is lower or higher in the actual world than the natural one, then this change can be attributed to human action.

To give an example (modelled on Christidis et al., 2013): scientists might be interested in extreme summer temperatures or heatwaves. Thus, they might compare how the July-mean temperature for a particular region has changed. To do so they would run a simulation of the 'natural world' for hundreds of years and observe the natural distribution of the July-mean temperature. They would then run a similar simulation with all the effects of human actions present and observe a different distribution of the July-mean temperature. Scientists would then focus on the extremes of these two distributions, that is, the area on the far end near the tails of the distribution where the temperatures are hottest or coldest. If the extremes in these two simulations are different, this difference can be attributed to human factors. This method attributes a *kind* of event, in this case extreme temperature, to human actions. But it does not attribute any *specific* observed event.

To attribute a specific event, scientists must take additional steps. They must relate the information they get about a kind of event from their simulations to a specific event that *has* occurred in the actual world. Christidis et al. (2013) used simulation results about July-mean temperatures in the Moscow region of Russia to analyse a 2010 heatwave. Since their data revealed that the actual 2010 July-mean temperature was about 25°C, scientists investigated the chance of exceeding a threshold of that value in both their 'natural' and 'actual' model simulations. They found that the event was indeed more likely given human influences, but it was an exceedingly rare event in both simulations. To declare that a specific weather event is attributable to human-caused climate change, one must have some standard for determining how large the increase in event likelihood must be for it to be deemed 'attributed'. Often that standard is a doubling of risk—i.e.

that human actions have resulted in an extreme event being twice as probable to occur—but there is no agreement between communities on this standard. While Christidis et al. did not employ this 'doubling of risk' standard in their study, the standard dates back to some of the earliest papers on extreme event attribution (see Stott et al., 2004). This, in a nutshell, is the attribution of a specific event.

Three reasons for event attribution

But why do it in the first place? As a National Academy of Sciences report points out: 'There is an element of scientific curiosity, but the primary motivation for event attribution goes beyond science' (NAS 2016: 21). Proponents tend to justify event attribution by claiming it will provide immense social benefit. First, event attribution is supposed to help societies adapt to the changes brought on by climate change. As Stott and Walton (2013: 277) point out: 'Societies may adapt poorly if they concentrate their efforts on developing resilience to recent extreme weather events that are set to become less, not more, likely under future climate change'. Second, proponents claim that event attribution can be an objective way of apportioning blame for the damage caused by extreme weather by separating deserving victims of climate change from undeserving non-victims. In that vein, Allen (2003) has explored holding entities responsible for the consequences of extreme weather through lawsuits, while and Thompson and Otto (2015) suggest that attribution can be used to identify victims and hold polluters responsible for damage caused. Lastly, extreme event attribution purportedly makes the effects of climate change visible in ways that may change public behaviour. Let us turn to analysing these three reasons for event attribution.

Reason 1: Event attribution enhances adaptive decision-making.

Response: But event attribution is backward looking; it cannot do much to help with adaptive decision-making with regards to future events.

If attributing extreme weather events significantly helped societies decide how to adapt to climate change, then the practice of attribution would be justified. However, there are good reasons to doubt that event attribution can make significant contributions in this area. Recall the research question that extreme event attribution addresses: 'How has the likelihood of an extreme event occurring in some particular geographical area changed due to human actions?' Now there are two ways to answer this question: one could say how the likelihood of a *type of extreme event*—say heatwaves in Europe in general—changed due to human actions; or one could say how the likelihood of a *singular specific event*—say the 2010 Moscow heatwave—changed due to human action. It is unclear how information about a singular specific event could help adaptive decision-making. City and regional planners do not plan for singular specific events. They want to know what general type of changes to expect, and the magnitude of these changes, so that they may prepare for events of that type. This information can be obtained from standard climate model projections. I therefore do not believe that adaptive-decision making can justify the attribution of specific events.

It is worth noting that both the type-level and specific-event attribution information is backward-looking and therefore is not directly relevant to adaptation planning. Event attribution provides information about changes in extreme events that have already occurred; it compares past or natural probabilities of event occurrence with current or actual probabilities of event occurrence. It is therefore not a projection or a prediction and thus does not provide any information about the future. Since adaptive planning is about anticipating future changes, it is unlikely that event attribution could provide the kind of significant benefits in this area that would justify its use or development (see Hulme, 2014 and Lusk, 2017, in **Further Reading** for more).

Reason 2: Event attribution can be used to hold polluters accountable for climate change damages due to specific extreme events.

Response: But event attribution does little to overcome the political challenges that arise when holding parties accountable for damages and it is not clear that it is reliable enough to play even a limited role in the process.

Specific extreme weather events are destructive. They are often responsible for the loss of infrastructure, crops, culture and people's lives. What if we could attribute these specific events to human-caused climate change? Linking the damages of a storm to climate change might enable us to hold those that caused climate change responsible for those damages. Many believe that extreme event attribution provides the necessary objective information that links extreme weather with human action to do just this.

However, using weather event attribution to hold entities accountable for climate change is not as straightforward as it would seem. Who is responsible for the human actions—the land use changes, the greenhouse gases and so on—that led to climate change? Well, certainly some countries, energy companies, car manufacturers, and individual consumers, among many other groups (see **Chapter 10**). But which of these groups should be held accountable when an attributable extreme event is identified? In many legal jurisdictions, in order to hold someone accountable you need to show a relatively direct link (in the USA, it's known as 'fair traceability') between the actions of the accused and the damage (see **Chapter 13**). But in the case of climate change that is very difficult because greenhouse gases mix in the atmosphere and build up over time. The actions that lead to climate change are collective; no single person, group of persons or company *causes* climate change. Different people, at different times, contribute differently to the problem. You might think that the Carbon Majors (Griffin, 2017)—essentially the top-100 fossil fuel users—could be held responsible given that their actions, collectively, are sufficient to cause climate change. Still, this ignores the interrelations between these companies and other companies or individual citizens that demanded energy, fuel, food and transportation. It also ignores the fact that governments (like the USA) subsidise and promote fossil fuel use.

The point is not that we can forgo assigning responsibility when it comes to the drivers of climate change; responsibility will have to be taken in order to solve the climate change problem. Rather, the point is that holding someone responsible for damages is a political act and a very difficult thing to do. Proponents often seem to think that the attribution of an individual extreme weather event provides the kind of objective information needed to escape this political quagmire. But it does little to help. For example, one needs to answer at the least the following questions: how reliable does our science need to be to lay blame? What should be done with damage from extreme events that cannot be directly linked to climate change? What groups should be held responsible? What responsibility do communities have to protect themselves? What portion of the damage should responsible parties cover (since they are not the only ones contributing to the problem)? These are all political—and legal—questions that would need to be answered about damage related to climate change. They don't become any easier to answer if we attribute individual storms to human action.

Even if we could answer these questions, it is not clear that the science of extreme event attribution is as reliable as it would need to be to lay blame. One major problem for scientists is that extreme events are by definition rare. Recall that scientists attribute an extreme event by comparing the frequency of event occurrence in the actual world with the frequency of occurrence in some (fictional) natural world. Since we don't have a natural world to observe, scientists use a proxy, often a computer simulation. To be reliable, we need to know that the computer model employed captures the frequency of extreme events reliably, both in the natural and in the actual worlds.

The difficulty for scientists—and they admit this—is that the real-world occurrence of many extreme events is so rare that just reproducing an observed extreme event frequency does not establish that the model is reliable. Scientists need to infer—from their knowledge of theory, the reproduction of more plentiful non-extreme events and the model's ability to simulate past climate change—that their models are reliable. How good this inference is, and therefore how reliable the attribution of an extreme event might be, is not entirely clear. It is very hard to quantify. This doesn't mean that scientists are wrong—there is always some uncertainty in science. But it raises the question of whether we have a good way of telling if the methods involved are as reliable as they need to be in order to hold groups responsible, for example in a court of law (see **Chapter 13**).

Furthermore, the extreme event attribution methodology I've described is not the only way to link human action and climate change. There is another approach, called the storyline approach (Trenberth et al., 2015). Roughly speaking, the storyline approach ignores probability of event occurrence and assesses how human factors influenced the characteristics, including the severity, of individual storms. Now you might think that the difference is a subtle one: both methods show how human actions influenced a dangerous event. A problem arises because the two different methodologies can give different answers to the question of human influence. Based on the change in occurrence an extreme event may not be influenced by human actions. But based on the storyline approach its severity might have been. Why should we favour one method over another? Anyone that wants to use event attribution to assign blame, or identify victims, needs to say why one method is used rather than another (or why both should be used, or neither). This is both a scientific question and a social question that remains unsettled (see Lloyd & Oreskes, 2018 in **Further Reading** for more on this point).

Reason 3: The attribution of specific events makes climate change visible when it otherwise wouldn't be.

Response: Maybe, but there are other ways to make climate change visible that are less confusing.

The human actions that influence extreme weather are the very same ones that produce climate change. Climate change is often thought of as abstract: no individual ever sees or feels it directly. They do, however, see and feel extreme weather. As the thinking goes, if individual storms can be attributed to climate change then seeing and feeling an individual storm just is seeing and feeling climate change (see Hulme, 2014, in **Further Reading** for more). When people can see and feel things—or at least, when they can be harmed by the things they see and feel—they are more likely to act.

The problem is that lay publics are bound to be confused by scientists' statements about the attribution of specific events. Why? Because if we are responsible with our reporting, then we have to say something like this: 'We have attributed this extreme event with a certain level of confidence to the human actions responsible for climate change, but remember that does not mean that human actions or climate change caused this particular extreme event'. Why do we need to say this? Because, as climate scientists admit, the extreme event—even when attributed to human actions—could have been completely natural in its causal origin, even if unlikely. The attribution of an extreme event simply shows that the *risk* of that event has doubled or surpassed some other threshold. It doesn't prove that human actions deterministically caused any *specific* singular event. Explaining how a single event could be blamed on anthropogenic climate change while simultaneously explaining that it could be natural in origin is difficult to do—and even more difficult to understand. If scientific misunderstanding can be avoided, then it should be avoided. Luckily, there is a better way to communicate with the public.

The following statement is much easier to understand: 'Our best scientific information indicates that the likelihood of an extreme event of this general type has increased due to the human actions that produce climate change.' This statement has many advantages. First, its framing is simple and it is consistent with the scientific evidence. Second, it is simple enough that an explanation of how probability and causality relate is unnecessary. And third, it does not require the attribution of a specific weather event. All this statement requires as evidence is that the likelihood of the generic type of event being considered has significantly changed.

Of course the payoff from this reframing of discussions of extreme events is that climate change can still be made visible. Not much is lost in talking about extreme weather events of a certain type rather than talking about a singular event. One might even speculate that by talking about event types, lay publics will become more comfortable with the responsible language of science. Or perhaps people in different parts of the world may see commonalities between the types of harm each are suffering due to climate change. This is all to show that there is more than one way to make climate change visible. The best ways do not seem to require the attribution of specific events to human actions.

Conclusion

As someone who strongly believes anthropogenic climate change is occurring and that substantial action should be taken to prevent its severe consequences, it is emotionally difficult to argue against the attribution of specific weather events. I want to be able to point at specific headline-grabbing weather events and blame their occurrence on human actions. Perhaps then—finally! —there would be a response on the scale necessary to remedy climate change. But I also know that armed only with our present knowledge doing this would be irresponsible—and likely have little effect. It would mislead the public into thinking that, without climate change, the damaging storm that they experienced or witnessed would not have occurred—something that may or may not be true. It would also do little to answer the massive political questions that continually result in apathy towards climate action. The attribution of specific events just doesn't have the purported social benefits that motivate its use and development.

What I also know is that there is a viable alternative: we can talk about types of events and responsibly claim that weather extremes of certain kinds have changed. We can say with confidence that the extreme events we witness are of the general kind of weather events that have become more or less likely. We can use this framing to help publics think in terms that are better aligned with extant evidence about climate change. Perhaps, then, societies will be willing to devote more time to the difficult political questions that stand in the way of effective action on climate change.

Further reading

Hulme, M. (2014) Attributing weather extremes to 'climate change': a review. *Progress in Physical Geography.* **38**(4): 499–511.

A review article which surveys the science of extreme weather event attribution in four stages: motivations for extreme weather attribution; methods of attribution; example case studies; and the politics of weather event attribution. It outlines some of the political challenges and dangers of using weather event attribution for guiding adaptation decisions, servicing the loss and damage agenda, or in the courts.

Lloyd, E.A. and Oreskes, N. (2018) Climate change attribution: when is it appropriate to accept new methods? *Earth's Future.* **6**(3): 311–325.

This article contrasts different methods of weather event attribution: probabilistic event attribution versus the storyline approach. The authors argue that while there is no 'right' or 'wrong' approach to weather event attribution, in different contexts societies may have a greater or lesser concern with errors of a particular type.

Lusk, G. (2017) The social utility of event attribution: liability, adaptation, and justice-based loss and damage. *Climatic Change.* **143**(1–2): 201–212.

Lusk traces the evolution of arguments for weather event attribution's social usefulness. He argues that probabilistic event attribution is unlikely to substantially contribute to litigation cases or to adaptation decisions, although it may potentially be relevant for addressing loss and damage.

National Academies of Sciences (2016) *Attribution of Extreme Weather Events in the Context of Climate Change.* Washington, DC: National Academies of Sciences, Engineering, and Medicine. Doi: 10.17226/21852.

This report from the US National Academy of Sciences examines the current state of the science of extreme weather attribution and identifies ways to move the science forward to improve attribution capabilities. It argues that as event attribution capabilities improve, they could help inform choices about assessing and managing risk and in guiding climate adaptation strategies.

Otto, F.E.L., van Oldenborgh, G.J., Eden, J., Stott, P.A., Karoly, D.J. and Allen, M.R. (2016) The attribution question. *Nature Climate Change.* **6**: 813–816.

A short survey of the current state of weather attribution science. The authors argue that every extreme event attribution study should clearly state the framing of the attribution question being asked. This should include whether conditional probabilities are being assessed or whether instead overall probabilities are being assessed.

Rudiak-Gould, P. (2013) 'We have seen it with our own eyes': why we disagree about climate change visibility. *Weather, Climate & Society.* **5**(2): 120–132.

Can the phenomenon of 'global climate change' be witnessed first-hand with the naked senses? This article contrasts the positions of 'visibilism' and 'invisibilism' as possible answers to the question, pointing out that how one answers this deceptively simple question has implications for the larger debate about scientific versus lay knowledge and the role of expertise in democratic societies.

Follow-up questions for use in student classes

1. Who gains from scientists attributing a particular weather extreme to human-caused factors? Are there costs from falsely attributing such weather extremes?
2. When climate scientists cannot provide information without relatively large uncertainties, should they refrain from answering questions in public at all?
3. How reliable does a scientific method need to be before we apply it to social problems? If the attribution of singular extreme events was somewhat likely to give false-positives or false-negatives, should we use it anyway?

4. Imagine two countries, A and B. Country A becomes rich after tapping a newly dis-
    covered vast oil deposit and, as it does, its culinary tastes change. Country
    B supplies country A with most of its food and would be economically devastated
    without this trade. As A's food tastes change, country B is forced to engage in envir-
    onmentally damaging agricultural practices to produce the new food A demands.
    These new agricultural practices make B more vulnerable to extreme weather. It just
    so happens B is then struck by a devastating extreme weather event. Should A be
    held responsible for the ensuing agricultural damage and loss of revenue suffered by
    country B? Does it matter if the extreme weather event was attributable to human
    action or not?

References

Allen, M. (2003) Liability for climate change. *Nature.* **421**(6926): 891–892.

Cheng, L., Hoerling, M., Smith, L. and Eischeid, J. (2018) Diagnosing human-induced dynamic and thermodynamic drivers of extreme rainfall. *Journal of Climate.* **31**(3): 1029–1051.

Christidis, N., Stott, P.A., Scaife, A.A., Arribas, A., Jones, G.S., Copsey, D., Knight, J.R. and Tennant, W.J. (2013) A new HadGEM3-A-based system for attribution of weather- and climate-related extreme events. *Journal of Climate.* **26**(9): 2756–2783.

Griffin, P. (2017) *CDP Carbon Majors Report 2017.* CDP Worldwide.

Hansen, G., Stone, D., Auffhammer, M., Huggel, C. and Cramer, W. (2016) Linking local impacts to changes in climate: a guide to attribution. *Regional Environmental Change.* **16**(2): 527–541.

Hasselmann, K. (1997) Multi-pattern fingerprint method for detection and attribution of climate change. *Climate Dynamics.* **13**(9): 601–611.

Herring, S.C., Christidis, N., Hoell, A., Kossin, J.P., Schreck, C.J. and Stott, P.A. (2018) Explaining extreme events of 2016 from a climate perspective. *Bulletin of the American Meteorological Society.* **99**(1): S1–157.

Jézéquel, A., Dépoues, V., Guillemot, H., Trolliet, M., Vanderlinden, J-P. and Yiou, P. (2018) Behind the veil of extreme event attribution. *Climatic Change.* **149**(3-4): 367–383.

Lehner, F., Deser, C., Simpson, I.R. and Terray, L. (2018) Attributing the U.S. Southwest's recent shift into drier conditions. *Geophysical Research Letters.* **45**(12): 6251–6261.

National Academies of Sciences (2016) *Attribution of Extreme Weather Events in the Context of Climate Change.* Washington, DC: National Academies of Sciences, Engineering, and Medicine. Doi: 10.17226/21852.

Otto, F.E.L., Coelho, C.A.S., King, A., Coughlan de Perez, E., Wada, Y., van Oldenborgh, G.J., Haarsma, R. et al. (2015) Water shortage in Southeast Brazil. *Bulletin of the American Meteorological Society.* **96**: 35–44.

Otto, F.E.L., Massey, N., van Oldenborgh, G.J., Jones, R.G. and Allen, M.R. (2012) Reconciling two approaches to attribution of the 2010 Russian heat wave. *Geophysical Research Letters.* **39**(4): L04702.

Risser, M.D. and Wehner, M.F. (2017) Attributable human-induced changes in the likelihood and magnitude of the observed extreme precipitation during Hurricane Harvey. *Geophysical Research Letters.* **44**(24): 12457–12464.

Shepherd, T.G. (2016) A common framework for approaches to extreme event attribution. *Current Climate Change Reports.* **2**(1): 28–38.

Stocker, T.F., Qin, D., Plattner, M., Allen, S.K., Boschung, J., Nauels, A., Xia, Y., Bex, V. and Midgley, P.M. (2013) IPCC, 2013: Summary for Policymakers. In *Climate Change 2013: The Physical Science Basis. Contribution of Working Group I to the Fifth Assessment Report of the Intergovernmental Panel on Climate Change.* Cambridge; New York: Cambridge University Press.

Stott, P.A., Stone, D.A. and Allen, M.R. (2004) Human contribution to the European heatwave of 2003. *Nature.* **432**(7017): 610–614.

Stott, P.A. and Walton, P. (2013) Attribution of climate-related events: understanding stakeholder needs. *Weather.* **68**(10): 274–279.

Thompson, A. and Otto, F.E.L. (2015) Ethical and normative implications of weather event attribution for policy discussions concerning loss and damage. *Climatic Change.* **133**(3): 439–451.

Trenberth, K.E., Fasullo, J.T. and Shepherd, T.G. (2015) Attribution of climate extreme events. *Nature Climate Change.* **5**(8): 725–730.

Uhe, P., Sjoukje, P., Sarah, K., Kasturi, S., Joyce, K., Emmah, M., van Oldenborgh, G.J. et al. (2017) Attributing drivers of the 2016 Kenyan drought. *International Journal of Climatology.* **38**(S1): e554–568.

4　Does climate change drive violence, conflict and human migration?

David D. Zhang, Qing Pei, Christiane Fröhlich and Tobias Ide

Summary of the debate

This debate engages with the controversy about whether climate change 'causes' human violence, conflict and migration. **David Zhang** and **Qing Pei** point to large numbers of quantitative studies which suggest significant correlations between climatic extremes and historical occurrences of violence, conflict and human migration. The key point, they argue, is that these correlations should be seen as causal when the studies are conducted over macro-scales of time and space. **Christiane Fröhlich** and **Tobias Ide** challenge these studies, and their relevance for the present-day, emphasising how conflict and migration are always multi-casual and usually driven by historical, social and political factors more important than climate. At the heart of this debate are questions about the scale of analysis and different interpretations of causality.

YES: Historically it does, over large scales of time and space (*David D. Zhang and Qing Pei*)

Introduction

In recent decades, scientists have found that many societal disasters and cultural collapses in human history were closely associated with climate change (Haug et al., 2003; Weiss et al., 1993). A great number of quantitative and qualitative studies concerning the relationship between climate change and human response in the past have been published. Our own work on China, Europe and the rest of the world found a strong historical association between climate change and war frequencies, migration events and population growth (Zhang et al., 2007). This association exists in different civilisations and countries ranging from the Middle Holocene (about 5,000 years ago) to near historical times. However, some social scientists and historians are reluctant to accept such an association between climate change and historical human crises. The association, they claim, is merely coincidental and is criticised for promoting **environmental determinism**.

Despite the debates raised by our work, and that of others, we believe that the findings regarding this relationship do not refute other existing theories about the causes of wars and human migration, or indeed about the complexity of other political changes. In this essay, we focus on the philosophical treatment of the subject of causality and on the importance of the temporal/spatial scale of research. We would like to correct the current misunderstanding about the relationship between climate change and human violence, conflict and migration. In brief, climate change should be viewed as *one of the factors*, along with others, leading to violence, conflict and human migration, rather than being the only factor to induce violence, conflict and migration in any given case study. However, climate change is a global phenomenon and so becomes the major factor to directly provoke large-scale natural ecosystem disasters and human economic turmoil, before consequently driving human ecological and societal chaos, such as war, migration, famine and population collapse (Zhang et al., 2007, 2011; and see Figure 4.1). The role of climate change should be understood as a first stimuli from the external natural environment to destabilise the normal functions of human societies in the past.

Our claim for a significant effect of climate change on past human societies is based on two major lines of reasoning: (1) causality analysis supported by many quantitative investigations; and (2) a macro-scale perspective across time and space. In our argument, rooted in empirical analysis, we prioritise the macro-scale and aggregate features of this relationship over the micro-scale and individual features. The causal analysis from a macro-scale perspective is not only theoretically sound, but also statistically proven. The causality and scale issues are clarified in the following sections as two fundamental concepts necessary to understand the significant role of climate change in driving violence, conflict and migration in human history.

Causal linkages

The first part of our argument is to emphasise the causal links between historical climate change and violence, conflict and human migration. Our claim for causation is based on the five criteria for the rigorous verification of causal links proposed by Haring and Lounsbury (1992). These criteria are as follows:

- a rational explanation of the relationship can be given;
- a strong association exists between the variables;
- a consistent association between the causal variable and the effect;
- the cause precedes the effect;
- the use of causal variable results in strong prediction.

The causal linkages argued for in this essay are not simply built according to one or even a few cases, but upon theoretical and statistical analysis of a large number of historical cases (i.e. large-N studies).

Whether climate change 'causes' human violence, conflict and migration is a subject of great debate among scholars of this field. In our earlier work we pioneered quantitative research on the relationship between climate change and armed conflicts in Chinese and world histories (Zhang et al., 2007, 2005). In recent years, a large number of other studies using quantitative methods have emerged and they also identified a strong association between climate change and societal crises generally at macro-scales during the pre-industrial era (Degroot, 2018; Tol and Wagner, 2010). Moving beyond the quantitative association, Zhang et al. (2011) attempted to verify whether such relationships are causal or not. They used datasets on European temperature change over the last 300 years and the historical records of proxies for European social, ecological and economic change as illustrated in Figure 4.1.

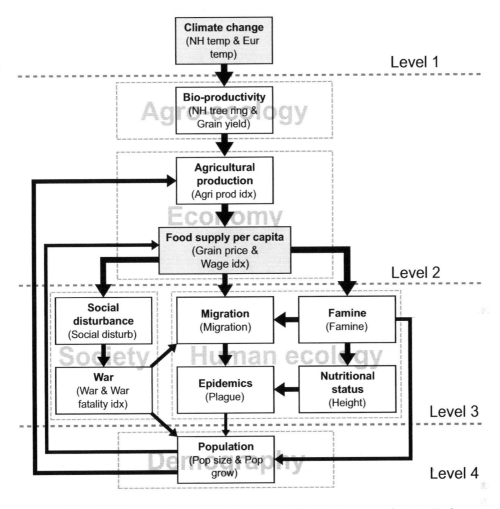

Figure 4.1 Causal linkages and cause/effect levels identified by European datasets. Each arrow points out the cause/effect relationships, which was verified by the five criteria. The black text represents the variables and the grey text represents proxies

Source: Zhang, D.D., Lee, H.F., Wang, C., Li, B., Pei, Q., Zhang, J. and An, Y. (2011) The causality analysis of climate change and large-scale human crisis. *Proceedings of the National Academy of Sciences*. 108(42): 17296–17301.

Causality is the relationship between a variable (the cause) and a second variable (the effect), where the second variable is understood as a direct consequence of the first. Zhang and his colleagues identified that climate change (Level 1 in Figure 4.1) caused great and widespread and impact on European societies through a domino effect (Zhang et al., 2011). First, climate change caused the failure of the agrarian economy (crop loss, the first effect; Level 2 in Figure 4.1). Such an effect then triggered a series of socio-ecological crises (the second effect; Level 3 in Figure 4.1), before eventually leading to the collapse of population (the third effect; Level 4 in Figure 4.1).

The causal relationship of each pairwise link in Figure 4.1 was tested and each met the above five criteria of causal verification. The links between these respective variables were well accepted as common knowledge by a large body of research in the natural and social sciences. For example, climate cooling leads to the failure of crop harvests and thus weakens the agrarian economy. This in turn brings about social unrest, migration, disease and famine, ultimately causing a decline in population size. According to this domino effect, seeded by climate change, we can confidently claim that climate change is the ultimate cause of human social, economic and ecological crises in pre-industrial Europe.

Our causality analysis, *undertaken at the macro-scale*, aims to identify a general account of climate change and human conflict. It offers an important philosophical paradigm. Such an approach is different from another school of philosophy which has influenced many researchers in the fields of social science and humanities, namely Mackie's so-called **'INUS' understanding of causality** (Mackie, 1974). Mackie stressed that events have a plurality of causes—that is, a certain effect can be brought about by a number of distinct clusters of causal factors. This assumption might well hold true when we approach a single case study, limited in time and space. For example, a specific human migration event may occur for a number of reasons. According to Mackie, each of these reasons is an 'Insufficient but Non-redundant' part of a condition which is itself 'Unnecessary but Sufficient' for the occurrence of the event (i.e. the INUS conditions). Critics denying the causal relationship between climate change and social crises often use a very specific example—such as the recent Syrian drought and civil war—to deny the linkage between climate change and a conflict or migration event. The key point to emphasise, however, is that our approach is rooted in a large number of cases, captured over long periods of time and on large spatial scales. It is here that we should look—and indeed find—common cause(s) and general association(s).

Mackie's paradigm does not attempt to look for common causes because he did not believe in their existence—although he did not deny the existence of statistical laws holding at certain scales. Statistical methods are adopted as the official mathematical language of many disciplines to construct causal relationships by inferring major connections (Pearl, 2000). Especially in the field of historical studies, some scholars attach great importance to statistical laws in the sense of patterns common to all periods and all types of social organisation. These patterns do not apply in every single case. But they do hold in a certain number of cases or when cases and data are aggregated at sufficient scale. We contend, therefore, that Mackie's philosophy does not apply to research that specifically seeks common causes of *large-scale and/or aggregated events*. His view on causality is also criticised by other philosophers on the basis that it 'fails to distinguish between genuine causes and mere joint effects of a common cause' (Psillos, 2009: 152).

Our argument that climate change drives violence, conflict and human migration is separate from saying that climate change is the cause of any individual social event, war or human disaster. This distinction serves as a reminder to pay attention to the question of scale—which we do in the next section, distinguishing between causality on a macro-scale from that on a micro-scale. Mackie's INUS causality paradigm might be appropriate for the latter, but not for the former.

The issue of scale

Geographic scale is a central conceptual issue of representation. Scale is closely linked to cartography because it strongly determines selection, simplification, classification and symbolisation. The issue of scale has been given serious attention by geographers for a long time. In recent decades, scale has attracted a growing academic focus in geography in two ways: (1) as a methodological issue inherent to observation and analysis; and (2) as an objective characteristic of complex interactions within and among social and natural processes (Sayre, 2005). In human geography, scale is raised by the foundational work of Taylor's classic 1982 paper (Dodds et al., 1997) with the 'three-scale structure' of model maps: the *micro-scale* of the urban mapped onto the domain of experience; the *meso-scale* of the nation state mapped onto the sphere of ideology; and the *macro-scale* of the global (Taylor, 1982). The importance of the scale issue requires any analyst to be mindful and reflective of the scale they select for analysis, because at different scales we may find different relationships between nature and society (Pei et al., 2016).

The five criteria for the verification of causal links (above) are most appropriately associated with the long-term and with large spatial scales. These scales enable researchers to include a large number of cases in the investigation of quantitative causal analysis. We argue that the long-term and the large spatial scale is necessary in the field of climate (environmental) studies to clearly reveal patterns of climate and society interaction. This contrasts with the narrative analysis of cause and effect, which we believe is more appropriate on smaller scales and to specific cases. This difference of approach, the different historical and geographical scales of research analysis, can partly explain disputes about the role of climate change in driving violence, conflict and migration in the human past. Although historians have sometimes considered changing climatic conditions in their micro-scale work, the factors *inside* the social system are usually credited with more explanatory significance than climatic factors in their studies of conflict.

There are three main reasons why such limitations prevail. First, climate change should be observed over the long-term and over large spatial scales. Case-based micro-scale studies of climate and conflict cannot fulfil this scale requirement. Social theories or phenomena might be able to explain the causes of individual events, but they will never be able to identify the causes of historical high peaks of war, conflict, migration and population collapses at large temporal/spatial scales. Nor will they be able to explain simultaneous occurrences of these peaks in different countries and continents. Such mismatches of scale will lessen the apparent significance of climate change in social relations and help explain why quantitative findings may be absent in such case studies.

Second, the impacts of climate change on human societies can only be discerned after a considerable lapse of time. As suggested by the IPCC, 30 years is the minimum time period to fully reveal the impacts of climate change. If the study period is not adequately long—in many of our studies we look over multiple decades or centuries—then the obtained time-lag effect of climate-conflict interactions will underestimate the impacts.

Third, unlike in the animal world, human society has buffering mechanisms (e.g. social and technological institutions) which can mitigate the influences of climate fluctuations. Although the influence of short-term climate variations on human relations can be relieved by these institutions, ensuring social resilience to the longer-lasting stimuli of large-scale climate change is much harder, especially in the past. Numerous studies have shown how such climatic effects have repeatedly triggered environmental disasters and societal collapses (e.g. Parker, 2013).

In contrast to the micro-scale historical and narrative approach, our macro-scale research in climate-society relations aims first to determine the general pattern, although it still acknowledges the complexity of human history and society. The macro-scale approach in historical geography hopes to generalise a common phenomenon based on a large number of cases which are usually collected over the long-term and at large spatial scales. In particular, in contrast to individual historical cases with their limitation of data availability, the macro-scale approach in time and space enables establishing big data sets for quantitative causality analysis. Macro-scale research in geography, with the objective of identifying general patterns, therefore always seeks a large-N sample. A micro-scale approach in history, to the contrary, treats each historical event as unique and particular.

The analogy of traffic accidents can help illuminate our argument. Traffic analysts and police officers can easily have different impressions of the cause of accidents due to different perspectives of scale. Traffic analysts attempt to identify the common cause of large numbers of accidents (e.g. patterns, frequencies, causal factors), whereas police officers tend to seek out the cause of each accident separately. For example, accidents increase dramatically in bad weather, such as during rainstorms, ice or snow. The analyst might well conclude that bad weather causes a high rate of accident. However, when police officers investigate these accidents, the cause can vary depending on each specific case, including drunk driving, an inexperienced driver or carelessness. The findings of both traffic analysts and police officers are correct according to their selected scale of analysis: macro-scale for traffic analysts and micro-scale for police officers. Other examples besides traffic accidents from other research fields emphasises our point. For example, medical researchers confirmed that smoking causes lung cancer through empirical studies using a very large number of cases. But not every lung cancer patient is a smoker, nor is every smoker a cancer patient. Macro and micro-scale perspectives give seemingly different, but in fact complementary, conclusions.

Conclusion

To summarise our argument: at a given spatial–temporal scale certain explanatory processes seem more fundamental than others in the system (Pei et al., 2016). Variation in the range of causes operating in any given instance may be considerable. Any geographically and temporally specific instance of violence, conflict or migration may not decisively be caused by climate change. Nevertheless, this possibility does not contradict our argument that climate change *does* cause violence and conflict when deriving a general picture obtained from all known historical cases at a macro-scale and over the long duration. Even if paradoxical, our argument does not refute other theories about the social or political origins of specific wars, violence and migration when examined at the micro-scale. Nor does it deny the complexity of political change in general historical processes.

The evidence upon which we have based our argument in this essay is gathered from studies that are distinguished from their predecessors and from case study research in terms of causal approach and temporal/spatial scale. Using statistical analysis and other quantitative methods, we argue that for most pre-industrial societies climate change is the ultimate trigger for large-scale human violence, conflicts and migrations. We also recognise, however, that if a society has strong buffering capacity and effective adaptive mechanisms, the impacts of climate change will be tempered and will probably cause less disruption. Our analysis rejects both environmental and social determinism, lines of thinking which emphasise only a single aspect of causation and which ignore the importance of scale dependency. The causal links between climate and conflict existing in the pre-industrial period may have been weakened or even broken by enhanced human adaptability in recent centuries, resulting mainly from industrialisation and other social and technological developments.

Nevertheless, human societies should not be over-confident when facing future potential human-caused and natural climate changes, not least because we have little idea about the magnitude of such changes to come. Following our argument, any natural/ social factor that causes widespread depletion of livelihood resources in industrial or post-industrial societies (e.g. global climate and environmental change, overpopulation, overconsumption or non-equitable distribution of resources) may lead to large-scale human crises due to the domino effect outlined in Figure 4.1. The extent of the associated crisis—in terms of violence, conflict and migration—will be largely contingent upon the degree of resource depletion and the effectiveness of advanced social buffering mechanisms. These latter include social institutions at international and national levels and a variety of adaptive social and technological developments.

NO: Other social, economic and political factors are nearly always more important (*Christiane Fröhlich and Tobias Ide*)

Introduction

The hypothetical connections between anthropogenic climate change, violent conflict and migration have garnered a lot of attention over the last decades, increasingly so with the effects of climate change becoming more and more visible. In the absence of adequate adaptation measures, the impacts of climate change such as changing droughts, storms, heatwaves, floods and sea-level rise are thought to create additional risks for livelihoods, especially among the poor and the politically marginalised. Livelihood insecurity is then assumed to stimulate migration to supposedly safer or wealthier places. This is especially the case if areas become permanently uninhabitable, for instance due to sea-level rise. Livelihood insecurity and resource scarcity are also hypothesised to result in higher inequalities and grievances and hence a higher risk of armed conflict. Furthermore, researchers are concerned that climate change facilitates a deterioration of the governance capacities of formal and informal institutions, potentially resulting in additional migration flows and conflicts.

Academic inquiry into such assumed causalities has questioned such predictions, as they often lack empirical support and a sound theoretical foundation. While this does not mean that anthropogenic climate change is irrelevant for future patterns of human migration or violent conflict, the connections are complex, multi-dimensional and extremely context-dependent. This essay will first outline the state of the art on climate-conflict and climate-migration links. Looking at the Syrian case, it will then demonstrate the difficulties of establishing causal links between (a) climate change and a specific drought; (b) drought and human mobility; and (c) human migration and conflict. By means of conclusion, we briefly discuss a few other cases, summarise the argument and flag the difficulties which arise from over-stressing assumed linkages between climate change, violent conflict and migration.

Climate, conflict and migration in a complex world: what does the literature say?

There are a large number of studies—both quantitative and qualitative—that find no impact of climate change on either migration or conflict. What is more, there are good theoretical arguments in support of such negative findings: floods and storms, for instance, often affect the mobility of troops and can destroy military equipment, hence making fighting less likely. Non-violent responses to the impact of climate changes (and the associated grievances) are considered more rational or culturally more appropriate in many cases because violence often induces a negative impact on social relations and economic assets. The literature on environmental peace-making and disaster diplomacy even argues that shared environmental challenges can provide entry points for positive-sum cooperation and reconciliation. Some pastoralist groups in the Horn of Africa, for example, set aside their differences during times of drought in order to engage in joint water and pasture management (Ide, 2018b).

When it comes to human migration, climate change can have a negative impact on economic growth and individual welfare. It is expected to lead to an increase in slow- and fast-onset weather events like prolonged droughts, heavier and more frequent storms, changing rainfall patterns as well as sea-level rise. Impacts include increased health hazards like a rise in malaria, the negative impact of heat waves on agriculture or the loss of houses and assets during storms. Depending on the individual positioning of a person, family or community, the likelihood of 'migration as adaptation' is assumed to increase (Black et al., 2011). This means that families or communities start to diversify their incomes, for instance by reducing agricultural activity in favour of other jobs, thereby potentially triggering rural-urban migration.

However, migration is an expensive activity and requires considerable financial, temporal and social resources to be successful. More often than not, people who are most affected by climate change are already marginalised and have little to no access to adaptive measures like changing to more climate-resilient crops or migration. What is more, the states most affected by climate change are often developing states in the Global South, where political make-up, economic structures and pre-existing conflicts may combine to form weak or fragile states. Importantly, colonial exploitation and the asymmetrical power relations between Global North and Global South in the current international system have contributed considerably to their continuous fragility. Such states have an even more difficult time adjusting to climate change impacts, so that global warming helps to produce 'trapped populations' (Zickgraf, 2018) who are willing, but unable, to migrate (for instance because they cannot pay traffickers). This illustrates that climate extremes can only catalyse what is already there, i.e. socio-economic marginalisation, political discrimination or state repression on the one hand, or weak democratic structures on the other. The 2011–2017 Californian drought and the 2018 European heat wave, for instance, sparked no violence and virtually no migration.

Also, migration movements in the context of climate change need to be related to pre-existing migration trends. Migration needs to be understood as a historically normal human activity, not a deviance from a supposedly sedentary norm (Bakewell, 2008). Research has established that the reasons for migration decisions are complex and strongly interconnected; therefore, we need to consider a diverse set of factors and explanatory frameworks in order to explain migration movements. Over-emphasising climate change as a driver of migration risks overlooking other, potentially more important factors.

To be fair, a good number of statistical, large-N studies still find a link between climate change and conflict or climate change and migration. However, in many of these studies the respective indicators of climate change have a rather low predictive power when compared with established drivers of conflict (e.g. pre-existing cleavages and low economic development) and migration (e.g. urbanisation and economic development). In other words, climate change is at best modestly catalysing existing conflict and migration trends rather than driving them. What is more, even advanced statistical research suffers from a number of inherent problems; its findings should be considered preliminary and interpreted with care. Even if past droughts or floods (which could be unrelated to human-induced climate change) catalysed conflict and migration, as proposed for instance for pre-industrial Europe and China, adaptation measures could avoid similar impacts in the future. Improved building structures and agricultural insurance systems are a case in point.

Furthermore, research on climate change, conflict and migration suffers from sampling and reporting biases (Adams et al., 2018). When it comes to climate-conflict links, for example, the literature is mostly focusing on countries where violence is already prevalent. Such a sampling on the dependent variable can distort results, among other reasons because it overlooks the many cases where climate extremes do not coincide with conflict (or migration), or where peaceful adaptation takes place. Furthermore, all conflict and several disaster databases extract their data from media and insurance company reports. In consequence, data on both conflicts and climate extremes in wealthy and accessible locations are well captured, while both are underreported in less 'visible' regions. This could result in artificially high correlations in statistical studies, especially when they analyse historical periods for which data quality and availability is limited (pre-1989 and especially pre-1946, not to speak of pre-industrial periods). Finally, even a significant and robust correlation between climate extremes and either conflict or migration does not necessarily imply causation. This is illustrated in the following section which discusses one particularly prominent example of such a correlation: the onset of the Syrian civil war after a pronounced drought.

Case study of Syria

The Syrian civil war of the last decade has been made out in public discourse as an example of the severe impacts anthropogenic climate change may have if its effects remain unmitigated and coincide with severe socio-political struggles (Gleick, 2014). Specifically, a prolonged drought period between 2006 and 2009 is considered (a) to have been caused by human-caused climate change; (b) to have resulted in a severe increase in internal, rural-to-urban migration and displacement in Syria; and (c) to have been an untold pre-story of the Syrian uprisings.

At the very least, the understanding of the relationship between climate change impacts and conflict via climate-related migration underlying such hypotheses contains the danger of implicitly over-emphasising some steps of the argument over others. First of all, attribution of a specific physical hazard such as drought to human-caused climate change is very difficult (see **Chapter 3**); the impacts of the hazard strongly depend on local political, social and economic conditions. Second, as mentioned above, migration is a complex phenomenon and environmental factors are just one of its many drivers. Environmentally motivated migration in Syria before 2011 happened at a much smaller scale than, for instance, labour migration to Lebanon. Third, we do not know how many people will leave their place of residence in the future due to human-caused climate change. The projected numbers differ considerably and are, in the end, hardly reliable as future adaptation measures are not factored in. Finally, orchestrating popular protest requires social networks built on trust and at least some kind of organisational structure. It remains unclear how new migrants, often living below the poverty line, could initiate large-scale, long-lasting popular uprisings, especially in repressive autocratic regimes like Syria.

We now apply these three general arguments to outline flaws in the climate-migration-conflict hypothesis for the case of Syria.

a.   *A human-induced drought?*

While there is little doubt that climate change is having real effects in Syria, it is difficult, if not impossible, to trace the causes of one specific drought back to human-caused climate change, at least with the scientific means available today. What is more, the drought in question here—a three-year period between 2006 and 2009—did not affect all parts of Syria equally; northeast Syria suffered much more than the rest of the country, with the three largest cities Aleppo, Damascus and Homs located elsewhere and actually receiving more rainfall than usual. Also, Syria was not the only state affected by the drought; in fact, northern Iraq was even more severely hit, but did not experience comparable social upheavals. Finally, the drought needs to be set in relation to other trends, because climate change is not the only force driving the decline and degradation of natural resources. For instance, the water resources available to Syrians have been severely decimated by overuse and pollution over the last decades. The current water deficit is about three billion cubic metres, which is about 20% of all water needs.[1] Other environmental stresses include deforestation, overgrazing and soil erosion, which may all be exacerbated by the effects of climate change, but are not caused by it (Selby et al., 2017: 234; Ide, 2018a).

b.   *Climate change driving migration?*

It is true that internal migration during the drought period 2006–2009 increased compared to periods of average precipitation. However, the number of people displaced was much lower than commonly assumed. The UN and the Syrian government estimated the total drought migration at 65,000 families, making it only the sixth largest reason for migration movement in Syria after population growth, the arrival of Iraqi refugees, general rural-to-urban migration, out-migration from Syria and the end of circular migration to Lebanon. What is more, these movements need to be seen in relation to other pre-existing migration trends, for instance seasonal agricultural labour migration or migration in the context of livestock herding (Selby et al., 2017: 239).

Most importantly, however, the transformation of Syria's economic system since 2000 needs to be taken into account as an exacerbating factor for migration dynamics in Syria. The turn towards a market economy, illustrated by the privatisation of state farms, the liberalisation of trade and the removal of key subsidies (fuel in 2008, fertiliser in 2009) led to a sharp increase in rural-to-urban migration. As droughts were nothing new in the region, and therefore crops like wheat were mostly irrigated, agricultural yields were less likely to be susceptible to drought than to subsidy cuts.

c.   *Climate migrants as protestors?*

As mentioned above, scholarly research provides little conclusive evidence for the hypothetical link between environmental migration and conflict (Brzoska and Fröhlich, 2016). On the contrary, migration is often seen as a way of adapting to climate change impacts

(Black et al., 2011), so that an increased number of migrants at a time of drought should not in itself lead to assumptions about these migrants' role in socio-political upheaval. Importantly, social movement theory suggests that migrants without prior connections to local (armed) groups lack the social as well as material resources to orchestrate protests of the scale and duration as the ones that started in Syria in 2011. Indeed, field research suggests that most migrants left when the anti-Assad protests started (rather than joining them) (Fröhlich, 2016). This does not preclude the possibility that they were later recruited by armed groups, but no convincing evidence on this has been provided so far (Ide, 2018a).

Conclusion

While the presumed climate-migration-conflict nexus in Syria receives considerable attention, similar arguments have been made for other cases as well. In countries like Ghana, Kenya, Mali and Nigeria, for example, less predictable rainfall patterns attributed to anthropogenic climate change facilitate the movement of pastoralists and their herds out of their 'core territories' and beyond traditional migration routes in search of water and pasture. This is supposed to stimulate conflict, for instance with farmers whose fields are destroyed by the animals, or with other pastoralist groups who are unwilling to share their pastures or compete for water resources. But such migration is triggered by a complex mix of factors, including land grabbing, economic marginalisation, the commercialisation of the cattle sector and a lack of state support. Similarly, the resulting conflicts almost exclusively occur in the context of pre-existing tensions. These are often fuelled by elites that strive to exploit violent unrest for their own political or economic agendas (for instance to get rid of political competitors or to secure access to valuable natural resources). In the absence of such factors, intergroup cooperation facilitating migration as adaptation is far more common as a response to (climate change-induced) resource scarcity (Ide, 2018a; Seter et al., 2018).

This evidence, along with the case of Syria and the quantitative literature discussed above, suggests that the impact of climate change on conflict and migration is complex, limited and strongly dependent on a range of (more relevant) contextual factors. In fact, basically all existing studies agree that political, economic and social factors are more important drivers of migration and violent conflicts than climate-induced environmental changes. There are no automatic causal linkages between human migration and conflict, as is illustrated by populations trapped by the impacts of armed conflicts or the secondary role migrants play in orchestrating protests (not to speak of violent insurgencies) (Fröhlich, 2016).

Importantly, academic and political debates about climate change, migration and conflict—while certainly necessary—can contribute to the (re-)production of misleading frameworks and counterproductive policy responses. In particular, because the impacts of climate change are felt most acutely in countries of the Global South, the reasons for these impacts *and* their solutions are often thought to be located within such regions. This runs the risk of overlooking the key role of non-climate related policies originating in the Global North in generating both first and second order impacts of climate change. In the Syrian case, for example, attributing the drought to climate change gave the Syrian government a way out of recognising its own role in exacerbating the effect the drought had on the Syrian population in the northern provinces (Selby et al., 2017). In a similar case, Sudan's former president al-Bashir named

Western carbon dioxide emissions and a climate-related drought as a key reason for the civil war onset in the Darfur region in 2003, while his regime played a major role in orchestrating the violence.

What is more, the existing power structures in today's international system also severely limit the ability of less wealthy regions of the world to adapt to and mitigate climate change impacts. We need to come up with a way of speaking about climate change which acknowledges the North-South divide and the need for climate justice (see **Chapter 11**), but without securitising global warming as a driver of massive population movements or violent conflicts. The latter discourse has been shown to lead to further walling-off policies and less public solidarity with those who are most affected by climate change.

Further reading

Baldwin, A. and Bettini, G. eds. (2017) *Life Adrift: Climate Change, Migration, Critique.* London/ New York: Rowman & Littlefield International.

This book contains a series of carefully selected essays which examine the causes and effects of human migration and how we treat those people labelled as 'migrants' and 'refugees'. The essays challenge emerging orthodoxies and stereotypes about climate change and human movement.

Hartmann, B. (2010) Rethinking climate refugees and climate conflict: rhetoric, reality and the politics of policy discourse. *Journal of International Development.* **22**(2): 233–246.

Hartmann critically examines the perceived threat of 'climate refugees' and 'climate conflict' in the context of alarmist rhetoric that has been deployed by a variety of actors, in particular US defence interests. She locates the ideological roots of these concepts in development theories and policy narratives about demographically induced migration, environmental refugees and environmental security.

Koubi, V. (2019) Climate change and conflict. *Annual Review of Political Science.* **22**(1): 18.11– 18.18.

Reviewing a wide range of evidence, this article concludes that a robust and general effect linking climate to conflict has not been detected. Nevertheless, substantial agreement exists that climatic changes contribute to conflict under *some* conditions and through *certain* pathways, for example in regions dependent on agriculture and in combination with other factors such as a low level of economic development and political marginalisation.

Parker, G. (2013) *Global Crisis: War, Climate Change and Catastrophe in the Seventeenth Century.* New Haven, CT: Yale University Press.

In this book Parker deploys scientific evidence concerning climate conditions of the period, and his use of 'natural' and 'human' archives to develop his thesis that changes in the prevailing weather patterns during the 1640s and 1650s caused great suffering and violence around the world.

Zhang, D.D., Pei, Q., Lee, H.F., Zhang, J., Chang, C.Q., Li, B., Li, J. and Zhang, X. (2015) The pulse of imperial China: a quantitative analysis of long-term geopolitical and climate cycles. *Global Ecology and Biogeography.* **24**(1): 87–96.

The long-term cyclical patterns of China's geopolitical shifts are of great interest and this study identifies the changing climatic and agroecological settings of these polities over time and analyses the relationships between climate change and historical geopolitical variations. This detailed empirical study finds that precipitation-induced ecological change was an important factor governing the macrogeopolitical cycles of imperial China.

Follow-up questions for use in student classes

1. Why are questions of scale so important when undertaking research on the connections between climate change and human behaviour?
2. Why is it difficult to attribute *specific* human migration movements to human-caused climate change?
3. What are the most important factors for the onset of violent conflict? Is climate change among them?
4. Review the supposed causal chain between human-caused climate change and the onset of the Syrian civil war. Which specific steps of this chain are more or less convincing?

Note

1 See: http://data.worldbank.org/indicator/ER.H2O.INTR.PC/countries/TN-LY-SY-1A-1W?display=graph Accessed 18 March 2019.

References

Adams, C., Ide, T., Barnett, J. and Detges, A. (2018) Sampling bias in climate–conflict research. *Nature Climate Change.* **8**(3): 200–203.

Bakewell, O. (2008) Keeping them in their place: the ambivalent relationship between development and migration in Africa. *Third World Quarterly.* **29**: 1341–1358.

Black, R., Bennett, S.R.G., Thomas, S.M. and Beddington, J.R. (2011) Climate change: migration as adaptation. *Nature.* **478**: 447.

Brzoska, M. and Fröhlich, C. (2016) Climate change, migration and violent conflict: Vulnerabilities, pathways and adaptation strategies. *Migration and Development.* **5**(2): 190–210.

Degroot, D. (2018) Climate change and society in the 15th to 18th centuries. *WIREs Climate Change.* **9**(3): e518. Doi: 10.1002/wcc.518.

Dodds, K., Smith, N. and Taylor, P.J. (1997) Classics in human geography revisited. *Progress in Human Geography.* **21**: 555–562.

Fröhlich, C. (2016) Climate migrants as protestors? Dispelling misconceptions about global environmental change in pre-revolutionary Syria. *Contemporary Levant.* **1**(1): 38–50.

Gleick, P.H. (2014) Water, drought, climate change, and conflict in Syria. *Weather, Climate, and Society.* **6**(3): 331–340.

Haring, L.L. and Lounsbury, J.F. (1992) *Introduction to Scientific Geographic Research.* Dubuque, IA: Wm. C. Brown Publishers.

Haug, G.H., Günther, D., Peterson, L.C., Sigman, D.M., Hughen, K.A. and Aeschlimann, B. (2003) Climate and the collapse of Maya Civilization. *Science.* **299**: 1731–1735.

Ide, T. (2018a) Climate war in the Middle East? Drought, the Syrian Civil War and the state of climate-conflict research. *Current Climate Change Reports.* **4**(4): 347–354.

Ide, T. (2018b) The impact of environmental cooperation on peacemaking: definitions, mechanisms, and empirical evidence. *International Studies Review.* **21**(3): 327–346.

Mackie, J.L. (1974) *The Cement of the Universe: A Study of Causation.* Oxford: Clarendon Press.

Parker, G. (2013) *Global Crisis: War, Climate Change and Catastrophe in the Seventeenth Century.* New Haven, CT: Yale University Press.

Pearl, J. (2000) *Causality: Models, Reasoning and Inference.* Cambridge: Cambridge University Press.

Pei, Q., Zhang, D.D. and Lee, H.F. (2016) Contextualizing human migration in different agro-ecological zones in ancient China. *Quaternary International.* **426**: 65–74.

Psillos, S. (2009) Regularity theories. In H. Beebee, C. Hitchcock and P. Menzies, eds. *The Oxford Handbook of Causation*. Oxford: Oxford University Press. pp.131–157.

Sayre, N.F. (2005) Ecological and geographical scale: parallels and potential for integration. *Progress in Human Geography*. **29**(3): 276–290.

Selby, J, Dahi, O.S., Fröhlich, C. and Hulme, M. (2017) Climate change and the Syrian Civil War revisited. *Political Geography*. **60**: 232–244.

Seter, H., Theisen, O.M. and Schilling, J. (2018) All about water and land? Resource-related conflicts in East and West Africa revisited. *GeoJournal*. **83**(1): 169–187.

Taylor, P.J. (1982) A materialist framework for political geography. *Transactions of the Institute of British Geographers*. **7**(1): 15–34.

Tol, R.S.J. and Wagner, S. (2010) Climate change and violent conflict in Europe over the last millennium. *Climate Change*. **99**(1–2): 65–79.

Weiss, H., Courty, M.-A., Wetterstrom, W., Guichard, F., Senior, L., Meadow, R. and Curnow, A. (1993) The genesis and collapse of third millennium north Mesopotamian civilization. *Science*. **261**(5124): 995–1004.

Zhang, D.D., Brecke, P., Lee, H.F., He, Y.Q. and Zhang, J. (2007) Global climate change, war, and population decline in recent human history. *Proceedings of the National Academy of Sciences*. **104**(49): 19214–19219.

Zhang, D.D., Lee, H.F., Wang, C., Li, B., Pei, Q., Zhang, J. and An, Y (2011) The causality analysis of climate change and large-scale human crisis. *Proceedings of the National Academy of Sciences*. **108**(42): 17296–17301.

Zhang, D., Jim, C., Lin, C., He, Y. and Lee, F. (2005) Climate change, social unrest and dynastic transition in ancient China. *Chinese Science Bulletin*. **50**(2): 137–144.

Zickgraf, C. (2018) Immobility. In R. McLeman and F. Gemenne, eds. *Routledge Handbook of Environmental Displacement and Migration*. Abingdon/New York: Routledge. pp. 71–84.

5 Can the social cost of carbon be calculated?

Reyer Gerlagh, Roweno Heijmans and Kozo Torasan Mayumi

Summary of the debate

This debate concerns whether and how one can use economic calculation—in particular the social cost of carbon—to guide decisions concerning climate change mitigation and adaptation. **Reyer Gerlagh** and **Roweno Heijmans** argue in support of calculating the social cost of carbon, claiming that it improves the effectiveness and efficiency of climate change decision-making and that it also helps to reveal normative disagreements between different political actors. They demonstrate a quick and simple way that the social cost of carbon can be calculated to reflect these disagreements. **Kozo Mayumi** takes a more critical stance to the whole field of cost-benefit analysis—within which the social cost of carbon sits—claiming that the nature and consequences of climate change undermine three key tenets of conventional economics: *scarcity*, *substitution* and *discounting*. For him, climate change is a '**wicked problem**' which requires political deliberation and ethical judgement, rather than economic calculation.

YES: The social cost of carbon is a simple and practical tool (*Reyer Gerlagh and Roweno Heijmans*)

Introduction

So important is the concept of the social cost of carbon (SCC) for policy that one of the founding fathers of this concept, William Nordhaus, was awarded the 2018 Nobel Prize in Economics for his pioneering work. The SCC measures the estimated costs to society, aggregated for the world and cumulated over the future, associated with one unit of carbon dioxide (CO_2) emissions. The SCC has also been the focal point of fierce debates amongst economists and also between climate policy advocates. We argue that just because these debates exist they should not be used as evidence against the usefulness of the SCC. On the contrary: the SCC helps to publicly and openly articulate the reasons for ethical and empirical disagreements about climate change policy. We start our defence of the SCC with three motivational reasons why we need a social cost of carbon.

Effective climate policy

Effective climate policy requires a carbon price as one of its central tools (see **Chapter 6**). Various alternative instruments are used by policymakers (Gollier and Tirole, 2015), such as subsidies for clean energy, information or nudges that bring about desired behavioural changes or the stimulation of clean technology innovation. All such alternatives share one common limitation: a failure to guide the economy towards a carbon-free future. Subsidies for clean energy increase demand for energy more than they reduce fossil fuel demand. Behavioural changes require an enormous amount of discipline from, and trust between, consumers, who would all need to forego individually preferable choices for the sake of the common good. And clean energy innovation only comes about when inventors and investors see a future market for their products. Without carbon prices, even subsidies cannot convince firms to put costly R&D expenses into developing clean solutions that will not pay back their inventors (see **Chapter 7**).

With carbon prices, on the other hand, the prospect of a big future market for low carbon technologies will 'pull-through' clean energy innovations. While alternative policy instruments falter, a carbon price targets its goal and enables the desired outcome. The reason is simple: better than alternative instruments, a carbon price signals the plain truth that every tonne of carbon emitted bears a cost to society. Though this simple insight is well established among economists, it is not so popular outside the discipline. But being economists, we are obliged to emphasise our conviction that firms and consumers respond to prices, whether they like it or not.

Clarifying trade-offs

To understand that a carbon price is important for the transition to a zero carbon economy is one thing. To set its level appropriately is quite another. The quest for a 'true' carbon price brings us to our second reason why calculating the SCC is both necessary and possible. Carbon prices are essential for understanding the trade-off between climate goals and other social objectives. We live in a world with overwhelming quantities of information, way in excess of what can be understood by any individual. Similarly, there are too many possible objectives to pursue—whether individual goals (be happy, find a partner, find a house, make a career) or societal ones (equal opportunities, social security)—to arrive at a rational prioritisation of goals. Policymakers and citizens alike therefore need assistance in understanding the compromises needed, or which choices we cannot afford. In this respect, the social cost of carbon represents a reversal of the classical image that ridicules economics as 'the dismal science' whose inhabitants can calculate prices without understanding values. This caricature does a gross injustice to economics.[1] Many economists fully comprehend the complexity of decision-making and the importance of balancing different values and trade-offs. For that reason they have devised 'price' as the ultimate instrument to guide complex decisions.

If we consider a social cost of carbon to be somewhere between €5 and €100 per tonne of CO_2 emitted (see the distribution of estimates shown in Figure 6 of van Den Bijgaart et al., 2016), and knowing that worldwide greenhouse gas (GHG) emissions currently amount to about 50 gigatonnes of CO_2 (equivalent) per year, we can immediately see that annual GHG emissions 'cost' society somewhere between €0.25 and €5 trillion. The high estimate exceeds 5% of world GDP. Replacing fossil fuels with renewable energy may be costly, but continuing with fossil fuels is *very* costly. The SCC therefore helpfully summarises valuable information that is indispensable for revealing the trade-offs between climate policies and other social goals.

The social cost of carbon as coordination device

This latter observation is also central to our third argument in support of an SCC: the social cost of carbon, implemented as a carbon price, acts as a policy coordination device. The most efficient distribution of efforts between countries and between sectors in tackling climate change is reached if every firm and consumer worldwide is confronted with the same carbon price (Pearce, 2003). This economic insight mirrors the scientific one that for the climate system it does not matter where a CO_2 molecule is emitted. The gas quickly mixes in the global atmosphere and contributes equally to climate change irrespective of its origin. Therefore, once citizens and policymakers accept GHG emissions cause climate change and that climate change causes damage, the social cost of carbon can act as a coordination device. Agreeing on the magnitude of this damage, which is what the social cost of carbon does, helps society limit climate change in the most effective way before it surpasses Earth's carrying capacity. One might feel intuitively repelled by the idea of putting a monetary value on the future of Earth, but economics teaches us to confront ourselves: when hard decisions need to be made we should not shun them, but instead seek a useful instrument to balance trade-offs between different options.

An example from another domain may help bring this insight home. Universal health care up to the maximum of technological possibility is desirable from an individual perspective. Anyone with a serious illness wishes treatment if it is available, whatever the cost. Yet if society were to satisfy *all* such individual demands, health care costs would spiral out of control. It is therefore clear that some selection needs to be made. The difficulty arises when health care has to select which treatments can, and which cannot, be afforded. Should we ration the most expensive cancer treatments, medication against Hepatitis C infections or open heart surgeries? And how about kidney transplants or prematurely born infants? These are hard choices to make. Given a limited budget, the decision to fund some treatments takes away the needed treatment for others. To assist, medics differentiate between those treatments that can be offered and those which are beyond budgetary capacities. The National Health Service in the UK has established a set of National Institute for Health and Care Excellence (NICE) guidelines. A treatment receives funding if it adds a quality-adjusted life year (QALY) for the patient at a cost below £30,000. Valuing life through a statistical approach sounds harsh, but it is the best operative protocol we can think of in the face of the hard truth that not all that is technically feasible is also economically, or politically, feasible.

The social cost of carbon is a tool comparable to the NICE guidelines. It guides us when deciding which of today's comforts or energy technologies we should give up or alter for the benefit of future generations. Equally, it helps us evaluate whether we should accept the sacrifice of the future for the sake of the needs and wants of present citizens. It thereby helps to coordinate our joint actions to get the best result out of our efforts.

Disagreeing about the social cost of carbon

An important factor throughout this exposition thus far has been the cost of future climate change. But how do we reach agreement on what that cost is? The social cost of carbon can help. This is because we do *not* need to agree on a single number; it suffices to agree merely on the concept. The different social costs of carbon then help to communicate the sources of scientific uncertainty *and* the sources of ethical disagreement in how we approach the answer. Only through transparent communication can we reach a common understanding, which is a prerequisite for joint action (Gerlagh and Liski 2017; Heal and Millner, 2014). Let us consider the two main sources of disagreement: (1) scientific and economic uncertainties; and (2) normative (i.e. ethical) disagreements.

The first class of uncertainties has gained much attention from the first generation of climate sceptics. When the fossil fuel industry began to understand the implications of climate science, they suggested scientific uncertainties implied that no policy action was the rational response. This argument is logically flawed. The mere fact that the likelihood of an unfavourable consequence is uncertain does not generally imply that no precautionary actions should be taken to avert it. Many people have home insurance or use a safety belt voluntarily. To insure what is valuable or to take preventive measures that reduce risks does not require proof that without these measures some loss will certainly be incurred. Similarly, knowing that increasing CO_2 emissions carries the risk of significant and worldwide damage to natural and human systems is sufficient reason to take preventive measures.

The truth is that scientists are not sure whether or not we are heading for severe climate change with temperature rises above 5°C and sea-level rise above two metres (see **Chapter 2**). But uncertainty is part of everyday life and it should come as no surprise that it is also part of climate change. As much as we can deal with uncertainty in our economies, so we can deal with it in climate policy. The SCC can be calculated for a scientifically pessimistic scenario, a scientifically optimistic scenario and for a so-called central case (as we will show below). It is then a matter of (political) preferences which weight should be given to each of these different values when making decisions.

Normative disagreement is different. The most common disagreement is between those who view climate policy as an economic investment decision versus others who see climate change from the perspective of future generations' equal rights (see **Chapter 11**). The first group wants to focus on the economic damages of climate change; they discount these future damages at the same rate as what they consider to be the standard discount rate for other investments, such as in machines, knowledge and infrastructure. The second group typically has a broader view of damages, including those affecting nature but which have no direct economic value. Moreover, they want to give equal weight to future affected people, poor and rich, currently born and unborn. As poor people have less to lose than rich people, their damages are small in strict economic terms. But as poor people also have less capacity to adapt, their damages are large in terms of human discomfort. Thus, those holding a narrower economic view will estimate lower costs than those with a broader, more ethics-based perspective. Finally, and not unimportantly, from an ethical point of view, when calculating the social cost of carbon future generations need not be discounted (i.e. valued less) compared to current generations. A very low (or even a zero) discount rate can be used (Cline, 1992; Stern, 2007).

The economic literature thus shows a big gap in the social cost of carbon calculated by 'descriptive' researchers who take economic markets as the source for their evaluations (e.g. Nordhaus, 2007) and by 'prescriptive' researchers who take ethical guidelines as the foundation of their quantitative assessment (e.g. Cline, 1992; Stern, 2007). However, this disagreement does not invalidate the concept of the social cost of carbon. To the contrary, the SCC can help clarify the differences between these two paradigms. Climate change is of global importance. Arguing with others using contradictory perspectives achieves little. A more constructive approach is to try and understand where and why views deviate and how to translate that insight into different preferences for policy action. The social cost of carbon enables this. Once we understand each other better, we can negotiate joint action.

How you can calculate the social cost of carbon

What better way to end an argument in favour of the social cost of carbon then to calculate it on the spot? Any undergraduate student in possession of the most basic numerical skills of adding and multiplication can do so. The calculations presented below are loosely based on van Den Bijgaart et al. (2016) and Rezai and Van der Ploeg (2016), who present a simple formula for the social cost of carbon. Importantly, they show their formula can almost perfectly reproduce outcomes of the most-used (and much more involved) integrated assessment model for climate-economy analysis: the DICE model (Nordhaus, 1993). We add to their basic mechanics some more recent insights as presented in Dietz and Venmans (2018).

We only need a few basic facts to understand climate change and the social cost of carbon. Scientific research suggests that the global climate response to accumulated CO_2 emissions is almost linear, almost immediate and almost permanent (cf. IPCC, 2014: Figure SPM.5). This is the surprisingly simple outcome of an intricate sequence of cause-effect chains that are too complicated to present here. The core insight is that for every trillion tonnes (teratonnes) of CO_2 emitted, the global average surface temperature increases by about 0.55°C (in fact somewhere between 0.4 and 0.7°C). Next, economists expect damages to be about 1% of global income (GDP) for every Celsius temperature increase (van Den Bijgaart et al., 2016). Yet these estimates vary widely, say between 0.1 and 3% of GDP, because there might even be some regional benefits from a warmer world that compensate for damages. The estimates are also dependent among other things on whether a narrow or a broad perspective of damages is taken.

Since world income as measured by GDP is about €70 trillion, we can immediately calculate the cost of annually emitted CO_2 as between €0.028 (0.4x0.001x70; the low case) and €1.47 (0.7x0.03x70; the high case) per tonne, with €0.38 per tonne (0.55x0.01x70) as a central case. This is an estimate of the annual damage to the world economy per tonne of CO_2 emitted.

We then only need to relate annual damages to what economists call 'the net present value' of damages. This is the cause of much controversy. It is related to the above-mentioned debate on discounting. Some authors argue that we should use market-based returns on investments as our guide (yielding a high discount rate). Others argue that ethical considerations should define our evaluation of future climate damages (yielding a low discount rate). When firms invest in capital, they require a payback time often within ten years. On the other hand, for big plants and infrastructural works, a longer payback time is acceptable. We suggest that in the context of big economy-wide structural changes caused by climate change a payback time of 30 years is a lower bound on the relevant economic time-window. Thus, the net present value would amount to 30 times the annual costs.

Taking the cumulative damages into account, the lower bound for the social cost of carbon is thus calculated to be €0.028 per tonne (low annual damage) times 30 years ≈ *€1 per tonne of CO_2*. Those who look at climate damages through an ethical lens would say that we should at least look 1000 years ahead and they also argue for a higher estimate of damages. For these people the social cost of carbon then is calculated as €1.47 per tonne (high annual damage) times 1000 years ≈ *€1500 per tonne of CO_2*. The central case would use, say, a 50 years payback time, yielding a calculation of €0.38 per tonne (median annual damage) times 50 years ≈ *€19 per ton CO_2*.

We have thus calculated a lower bound (€1), upper bound (€1500) and central value (€19) for the SCC. For economic decisions, the average often provides the relevant information. If we allow some simplification, we can say that because the lower bound is so small it is virtually zero. Thus the average is mainly driven by the probability of the central estimate versus the probability of the high estimate. If we believe the central case has 50% likelihood and the high estimate has a 1% likelihood, this would warrant a carbon price of €9.5 (€19 x 0.5) + €15 (€1500 x 0.01) ≈ *€25 per tonne of CO_2.* Alternatively, we might split the 50% likelihood for the central estimate into a 40% chance for €19 per tonne and a 10% likelihood that the correct estimate is high, but not extreme, say about €150 per tonne. If we then add the 1% likelihood for the very high end of €1500, that would bring the expected social cost of carbon to be 0.4x19 + 0.1x150 + 0.01x1500 ≈ *€37 per tonne of CO_2.* Note that this value substantially exceeds the central estimate of €19 per tonne.

How do these values relate to typical estimates for the SCC that we can find in the literature? Indeed, the central estimate we are suggesting here (i.e. €19) is in the range we find in the scientific literature based on 'best guesses', while studies that consider uncertainties for the underlying parameters indeed find higher values, in the same way that our average (€25–37) exceeds our 'best guess' (€19) (van Den Bijgaart et al., 2016). That is, the back-of-the-envelope calculation we illustrate here represents much of the variation for the SCC found in the literature.

Our simple illustration suggests a lower bound for the social cost of carbon of €1 per tonne, an upper bound of €1500 per tonne and a 'most likely' value of €19 per tonne. This may seem to represent a huge range of uncertainty, but importantly, we can see clearly where the differences in these estimates come from. The *lower bound* is based on an optimistic perspective of large-scale social adaptation to climate change and low damages and on comparing climate policy with a typical firm's decision horizon for a plant with 30-year lifetime. The *upper bound* is based on a more pessimistic perspective where the poor suffer most from climate change and their costs are inflated to make them count as though they were rich people. The upper bound also assumes climate policy should be considered an ethical decision where we must look at least 1000 years ahead. The central estimate steers a middle-of-the-road course.

Conclusion

There is no need to quarrel about the social cost of carbon. We share one Earth and we share it with our children. As much as we raise our families in accordance with our own values, so too we may keep our different views on climate change and our different convictions about which people should take responsibility for sorting it out. The social cost of carbon is a simple and practical tool to convert these beliefs into a convenient numerical index. When given the basic insights everyone can calculate it on the back of an envelope. Not only does the social cost of carbon help us to comprehend differences in practical and ethical reasoning, it can also help in the search for common ground between different people and nations for the practical policies that will determine our shared future.

NO: There are fundamental problems with cost-benefit analysis when applied to climate change (*Kozo Torasan Mayumi*)

Introduction

The social cost of carbon (SCC) is a monetary measure representing the long-term social cost of a unit of CO_2 emitted into the atmosphere. The SCC approach—usually represented in terms of the discounted marginal cost of climate change damage—falls within the framework of cost-benefit analysis (CBA). For the case of climate change, CBA seeks to offer a formal monetary framework for use in the evaluation of adaptation and mitigation policy options. CBA provides decision-makers and concerned publics with a guide for future collective discourse and policy formation. However, if the Fifth Assessment Report by the IPCC is credible, applying CBA to climate change is problematic. This essay argues that CBA and the social cost of carbon should not be applied to climate change.

I develop my argument by discussing the three fundamental tenets of *scarcity*, *substitution* and *discounting* that form the theoretical edifice of CBA. The essay argues that applying scarcity, substitution and discounting to climate change exceeds the legitimate scope of CBA within which these concepts were originally formulated. Climate change is a problem characterised as post-normal science (Funtowicz and Ravetz, 1993)—where decision-making is urgent, stakes are high, values are in dispute and knowledge is uncertain. I close with a brief discussion of an alternative approach to tackling this post-normal '**wicked problem**' of climate change, one that is more suited than CBA and the social cost of carbon.

Essential points from the IPCC report

There are four important conclusions from the IPCC Fifth Assessment Report (IPCC, 2014) which are crucial for my critique of the social cost of carbon. First, anthropogenic climate change is pervasive and irreversible on a multi-century to millennial time scale unless there is rapid and sustained reduction in greenhouse gas (GHG) emissions. Second, many regions will experience changes in climate related extremes— such as heatwaves, droughts, floods, cyclones and wildfires. Such climate variability presents significant dangers to ecosystems and livelihoods, particularly in developing countries with low incomes. Third, marine life such as coral reefs face increased ocean acidification and higher ocean temperatures which threaten their long-term viability. Global redistribution of marine species and reduction of marine biodiversity in sensitive regions will endanger fisheries that rely on healthy ecosystems.

Fourth, urban areas are subjected to increased future risks affecting people, assets, economies and ecosystems. These risks result from heat stress, storms and extreme precipitation, inland and coastal flooding and landslides. Rural areas will likely experience major impacts on water availability and supply, food security, infrastructure and agricultural incomes, including shifts in the production areas of food and non-food crops. Importantly, climate change seriously threatens Net Primary Production (NPP) (Vitousek et al., 1986), the most important biophysical basis for biological life on the Earth and which ultimately regulates economic production from agriculture, fishery and forestry. NPP is non-substitutable and yet human life depends upon it.

Reconsidering scarcity, substitution and discounting in conventional economics

Scarcity

The conventional economic framework which underpins the social cost of carbon dates back to the socioeconomic conditions prevailing in the UK in the nineteenth century during the Industrial Revolution. At that time, people started to anticipate that perpetual economic growth could be maintained by a combination of abundant fossil fuels and expanding economic production. Since then, however, limitless desire for goods and money in anticipation of perpetual growth has become the foundation of the *scarcity* concept in conventional economics.

Scarcity refers to the situation in which a persistent gap exists between available goods and limitless wants. The modern world is full of economic goods that allow a *reasonable lifestyle*, but there are not enough of them to satisfy *limitless wants* (Samuelson and Nordhaus, 2010). Hubin's term 'moderate scarcity' (Hubin, 1989), exactly corresponds to the definition of scarcity given by Samuelson and Nordhaus. The essence of the scarcity concept appears in Arrow and Debreu's classic article 'Existence of an equilibrium for a competitive economy' (Arrow and Debreu, 1954). Exactly parallel to the idea of moderate scarcity in conventional economics, they assumed that 'every individual could consume out of his initial stock in some feasible way and still have a positive amount of each commodity available for trading in the market' (p.270).

In reality, and in contrast to meeting unlimited wants, an economy must make the best use of its *sufficient*, yet limited, available resources—i.e. it must operate under the condition of moderate scarcity. This therefore leads to the efficiency criterion. This efficiency criterion uses CBA to choose investment projects that maximise the present monetary value of total net economic benefits, benefits accruing over a certain period of time and with a certain discount rate. The scarcity concept as used in conventional economics does not therefore apply to the condition of 'severe scarcity', a condition under which it is very difficult to satisfy basic needs or to obtain the necessary provisions for subsistence.

The Contingent Valuation Method (CVM), often used in CBA, asks people how much money they are willing to pay for protecting the provision of goods coming from ecosystems. In the case of 'severe scarcity', however, the market mechanism cannot give all individuals an acceptable share of goods. Here, the most important question to ask is rather what type of subsistence goods are necessary for a decent life. Poor people in developing countries often lack purchasing power, so the monetary value they report in CVM is necessarily too small, ultimately leading to ruining their subsistence resources for survival. Yet if—as the IPCC report suggests—irreversible climate change threatens NPP, 'severe scarcity' of economic goods must eventually occur. The situation envisioned by conventional economics and assumed by CBA—that of 'moderate scarcity'— does not apply in the case of climate change, where conditions of 'severe scarcity' will unfold.

Substitution

In conventional economics, smooth substitution among goods and production factors— one of the pillars of CBA—is essential for price mechanisms to work. According to Marshall (1920), quasi-constancy of the marginal utility of money—the change in the satisfaction or the benefit, called 'utility', resulting from an increase in additional money spent—is assumed to be compatible with a society of 'middle-class individuals'. In such a society, a substantial part of individual income is spent on numerous mere conveniences. In relation to total income in middle or high income societies, most mere conveniences involve marginal expenditures. A slight income variation causes one or more such conveniences to *disappear* from the individual's budget (when income goes down) or to *appear* as one or more new entries in the budget (when income goes up). In such conditions it is reasonable to assume that the marginal utility of money for all conveniences is almost the same because individuals are indifferent to buying any one particular convenience. Consequently, substitution between these convenience goods is said to be 'smooth'.

On the other hand, substitution in the production process is much less smooth. Industrial production processes are conditioned by the physical properties of material objects for particular purposes, so complete substitution of one material for another usually requires a considerable period of time. A typical example used to illustrate this from the nineteenth century is the substitution of wrought iron with steel, in which the latter was generated through open hearth furnaces where excess carbon and other impurities were burnt out of pig iron. But in conventional economics, substitution is treated as if there were no essential difference between individual consumer choice and industrial production. In particular, the substitution of primary energy sources in economic production (e.g. renewable energy for fossil fuels) is the key element in mitigating climate change (see **Chapter 7**). It is striking that IPCC does not seriously investigate the possibility of substituting primary energy sources to replace fossil fuels and thereby reduce GHG emissions. In its Fifth Assessment Report (2014), of the 1,454 pages of the Working Group III section on mitigating climate change only three pages were devoted to fossil fuels, renewable energy and nuclear energy.

There are formidable barriers to achieving the smooth substitution of primary energy sources that is assumed within the theoretical basis of CBA. Three of these barriers are: the heavy dependence on incumbent fossil fuels to generate electricity; the insufficient reserves of uranium-235 and the difficulty in creating commercial nuclear breeder reactors; and the limited supply of silver for multi-crystalline silicon wafer-based solar cells in the photovoltaic (PV) systems. These three barriers are discussed briefly in turn.

Although electricity is considered to be the cleanest form of energy, its generation still depends heavily on fossil fuels with two thirds of the world's electricity in 2015 being generated this way (Table 5.1). To generate sufficient electricity without using fossil fuels is crucial for mitigating climate change. Unfortunately coal—the most intensive source of GHG emissions—still contributes almost 40% of global electricity generation. Thanks to shale oil gas production, the USA reduced its coal-fired electricity generation from 52% in 2008 to 34% in 2015. However, this reduction is likely temporary. Along with the USA—and despite government claims—China and India remain heavily dependent on coal-fired electricity generation. There remains little evidence of smooth substitution away from coal-based electricity.

In relation to nuclear energy, although the large majority of electricity generation in a country like France comes from nuclear power generation, the total uranium-235 supply does not satisfy current world demand. The proven reserves of uranium-235 are surprisingly limited at about 7.6 million tonnes (2015 estimate). If nuclear energy were to supply the total world primary energy consumption as of 2008, the uranium-235 reserve would last less than eight years (Nuclear Energy Agency, 2016). Under these limiting circumstances, supporters of nuclear power generation have been trying to establish a fast breeder reactor that uses a mixed oxide consisting of plutonium dioxide and uranium dioxide. But developing a commercial fast breeder reactor involves four phases, progressing from experimental reactors, to prototype and demonstration reactors, to full commercial reactors. Japan, for example, only reached the prototype stage of such development before abandoning plans for the Manju breeder reactor and started its

Table 5.1 Percentage share of primary energy sources for electricity generation in 2008 and 2015

	Coal		Natural gas		Hydroelectric		Nuclear		Oil		Other renewable	
	2015	change	2015	change	2015	change	2015	change	2015	change	2015	change
USA	34	-18	32	+14	6	+4	19	-4	1	0	7	+4
France	2	-2	4	0	10	+6	78	-6	0	-2	6	+4
China	70	-19	2	+1	19	+12	3	+1	0	-1	5	+5
India	83	+1	5	-2	1	-3	3	+1	2	-2	6	+5
Japan	34	+6	40	+16	8	+5	0	-30	10	-2	8	+5
World	39	-6	23	+2	16	+10	11	-5	4	-2	7	+2

Source: Data compiled from EDMC (2011) and EDMC (2018), Handbook of Energy and Economic Statistics, Energy Conservation Center, Tokyo.

decommissioning in 2016. Overcoming barriers to energy substitution using nuclear technology seems unlikely in the foreseeable future.

A third barrier to smooth substitution in the energy sector concerns solar PV. Direct use of solar energy such as in PV systems is considered to be a good candidate for a globally available renewable primary energy source to substitute for fossil fuel. PV installations based on first generation multi-crystalline silicon wafer-based solar cells represent the most widely adopted technology worldwide, with a market share of about 95% (Fraunhofer Institute for Solar Energy Systems, 2018). Indeed, despite development of second generation (thin-film) and third generation solar cells, the share of crystalline silicon wafer-based solar cells holds firm with no apparent sign of decline. Silver is used in a specialised paste for the contact metallisation of silicon wafer-based cells. Although the decrease of silver consumption per cell has been remarkable in recent years, world silver reserves are limited. A solar PV deployment to provide just 30% of the current global electricity demand would consume one third of the currently estimated world silver reserves (Lo Piano and Mayumi, 2017). The silver requirement for a large-scale electricity generation by PV systems is a formidable problem. The principle of smooth substitution that is central to the application of CBA is again to be questioned.

Discounting

A third tenet of CBA, and hence of the calculation of the social cost of carbon, is the notion of discounting. It is customary to calculate the present monetary value of an investment without considering the origin or justification of discounting. Unlike all other material objects, money defies the first and the second laws of thermodynamics. Defying the *first* law, money can be created out of nothing by national banks and financial agents. Defying the *second* law, money does not decay functionally, even though the physical structure of a coin or a bank note decays.

Every material object has a material structure—i.e. a *structural component*—and every material object has a particular purpose for use—i.e. a *functional component*. As the structural component of a material object decays due to the second law of thermodynamics, its functional component also decays. As a car rusts it becomes unusable. Material objects become unsuitable for the particular purposes for which they were originally intended (Mayumi and Giampietro, 2018). However, money is different. Even if its physical structure decays, money still functions as its originally intended purpose. As USA law stipulates:

> Lawfully held mutilated paper currency of the United States may be submitted for examination in accord with the provisions in this subpart. Such a currency may be redeemed at face value if sufficient remnants of any relevant security feature and clearly more than one-half of the original note remains.
>
> (Legal Information Institute, 2017)

A decayed banknote still functions as a banknote. A decayed car does not still function as a car.

Because the function of money does not change over time, a qualitative gap opens up between the value of money and the value of goods: money maintains its original function, while material objects lose theirs. Money grows quantitatively over time through a positive interest rate, whereas the value of material goods depreciates. Discounting monetary value is therefore justified in conventional economics and the superiority of money over material goods is supported institutionally and legally. The owner of money can then in principle dictate the timing of transactions with people who have to sell goods in order to limit the structural decay of those goods.

Furthermore, money has a dual nature that is not always recognised by conventional economists. On the one hand, money is recognised as an indicator of individual wealth. Yet money *also* represents a debt for the whole community because it entails a promise to pay 'the bearer', whether this repayment is in terms of existing goods or on the promise of future goods. The greater the accumulation of money, the greater the long-term biophysical debt of the whole community. Other financial assets also have such dual nature. At some point, indefinite future economic growth fuelled by money expansion will be confronted by biophysical constraints (Mayumi and Giampietro, 2018), yet conventional economists do not recognise this dangerous aspect of money expansion.

Discounting monetary value from an individual perspective may make sense, but it is not suitable for facing climate change where collective decisions are necessary. Maximising the present monetary value, as is assumed in CBA, represents only an individual perspective on money and consumption. But as Bromley (1990: 97) properly states, 'it is a value judgement for the economist to claim that economic efficiency *ought to be* the decision rule for collective action'. W.S. Jevons for example—one of the founders of conventional economics in the nineteenth century—explicitly stated that discounting should not take place (Jevons, 1965). And the Nobel Prize-winning American economist Joseph Stiglitz has declared that conventional economic analysis is really only concerned with answering questions over 'the next 50–60 years' (Stiglitz, 1997: 269).

Despite these warnings, the power of discounting within CBA is remarkable and discounting is central to the calculation of the SCC. Over a time horizon of 50 years and a 1% discount rate, the present value of €1 decreases to €0.61; at a 5% discount rate it decreases to €0.087; and with a 10% discount rate, after 50 years €1 becomes worth just €0.0085. Discounting as used in the calculation of the social cost of carbon makes no sense, since the time scale of climate change is so long—a hundred years or more and much longer than Stiglitz's 50–60 years.

An alternative way to deal with climate change

The risks of climate change have a number of unique features that cannot adequately be captured by the calculation of its economic externality, i.e. the social cost of carbon. This is so whether the social cost of carbon is calculated as either €1, €19 or €1,500 per tonne of CO_2 emitted—as done by Gerlagh and Heijmans earlier in this chapter. The real challenge of climate change does not lie in calculating the social cost of carbon in monetary terms. It lies in considering how to distribute the related inter-generational and intra-generational burdens and gains that result from climate change in a sustainable and equitable way (see **Chapter 1**). This cannot be done effectively using the tools of CBA and the social cost of carbon.

As explained by Tainter et al. (2015), and many others, there is a more modest and practical approach to dealing with sustainability problems such as climate change which avoids CBA and the social cost of carbon. This approach asks four questions: what is to be sustained? Sustained for whom? Sustained for how long? And sustained with what kinds of gains and sacrifices? Based on such an approach, the following four objectives are necessary for dealing sustainably with climate change:

- reaching a constructive agreement through deliberation and cooperation on what should be sustained;
- discussing seriously the local, regional and global distribution issues related to intra-generational and inter-generational equity;
- educating ourselves in such discussions to look much further ahead than in most current economic analysis;
- itemising, as far as possible in non-monetary (i.e. ethical) terms, the gains and sacrifices that are achievable in a hierarchically organised world in which socioeconomic conditions are skewed.

Conclusion

I have based my argument on three points. First, scarcity in conventional economics cannot be applied to situations such as climate change where 'severe scarcity' of goods prevails. Using the social cost of carbon to deliver an efficient allocation of resources under such conditions cannot be achieved. Second, in order to mitigate climate change alternative primary energy sources to fossil fuels are essential yet, as shown above, nuclear energy and PV systems are not hopeful candidates. In other words, the smooth *substitution* of energy generation assumed in CBA cannot be guaranteed. Third, *discounting* the future value of money may be justified for an individual's monetary decisions, but is a questionable practice for collective decisions dealing with long-term climate change. Applying the criterion of maximising present monetary value systematically underestimates the biophysical requirements of future generations, thereby offering a false prospectus of what are, in the end, exhaustible resources.

It is instructive to conclude this essay with the following consideration. William Nordhaus—one of the founding fathers of the SCC and a Nobel Prize Winner in economics—once presented an equation for calculating the cost of climate change (Nordhaus, 1992: 1316):

$$d(t) = 0.0133 \left[\frac{T(t)}{3} \right]^2$$

where *d(t)* is the fractional loss of global output from greenhouse gas warming in period *t;* and *T(t)* is the temperature in period *t*, usually measured in degrees kelvin (K).

Because *d(t)* is a dimensionless number, the relation does not make sense unless appropriate dimensions are assigned to the two constants in the equation: 0.0133 and 1/3. Otherwise, it becomes merely an arbitrary curve-fitting exercise. It is depressing to see that the procedure adopted by this prize-winning economist to calculate the social cost of carbon is nothing but a statistical convenience, completely ignoring the dimensions of the variables being related to each other. In my view, applying the social cost of carbon to climate change results is formalised nonsense (Mayumi and Giampietro, 2012).

Acknowledgements

I would like to express sincere thanks to Donald Sturge of Tokushima University for help in improving the language in this essay. I would like to express my gratitude to my friends, Herman Daly, Silvio Funtowicz, Mario Giampietro, Tiziano Gomiero, John Gowdy, Shunsuke Managi, Shigeru Matsumoto, Giuseppe Munda, John Polimeni, Jesus Ramos-Martin,

Jeroen van der Sluijs, Joseph Tainter and Hiroki Tanikawa who gave useful comments, moral support and reference materials. I also appreciate the help of editor Professor Mike Hulme at every stage of preparing this essay. Remaining errors are all mine.

Further reading

Gerlagh, R. and Liski, M. (2017) Consistent climate policies. *Journal of the European Economic Association.* **16**(1): 1–44.

The conceptual gap between the ethical and the descriptive view of the SCC has all the features of a dividing line between two opposing factions. Most researchers stick to just one position. This article by Gerlagh and Liski is an exception, since they seek to integrate both views and offer a generic discussion of social choice under value disagreement. This paper provides a bridge between cynical economists and romantic ecologists.

Gowdy, J.M. (2013) Valuing nature for climate change policy: from discounting the future to truly social deliberation. In R. Fouquet, ed. *Handbook on Energy and Climate Change.* Cheltenham: Edward Elgar.

Gowdy's chapter in this handbook argues that the magnitude, suddenness and long-term consequences of climate change call for a radically new approach to valuing nature. Rather than relying on market-based individual choice, it outlines an approach that relies on social discursive processes for collectively solving the problem of intergenerational sustainability.

Pearce, D. (2003) The social cost of carbon and its policy implications. *Oxford Review of Economic Policy.* **19**(3): 362–384.

In this article David Pearce explains the meaning of the SCC concept, outlines the different views on how to cumulate future damages into a present estimate of economic damage and argues the need for a uniform pricing of carbon through the economy. Because the calculated SCC tends to increase over time—similar, say, to house prices—the typical estimates of the SCC in the literature have increased considerably since this article was published in 2003.

Pezzey, J.C.V. (2018) Why the social cost of carbon will always be disputed. *WIREs Climate Change.* **10**(1); e558. DOI: 10.1002/wcc.558.

This nicely argued opinion essay offers a critical view of the SCC, partly because of the deep and irreducible uncertainties involved in its calculation, but also because it feeds an illusion of the optimal allocation of resources over a centennial time-scale.

van Den Bijgaart, I., Gerlagh, R. and Liski, M. (2016). A simple formula for the social cost of carbon. *Journal of Environmental Economics and Management.* **77**: 75–94.

The simple formula for the SCC presented at the end of this essay is inspired by this article. It provides an extensive motivation for their derivation of the SCC and adds an extensive analysis to support the validity of their rule compared to calculations based on fully functioning integrated assessment models.

Follow-up questions for use in student classes

1. In the context of the social cost of carbon what does 'social' and 'cost' mean to you?
2. Do you agree that society cannot afford to provide *all* available medication to patients, even if such medication can save lives? What is your view on the NICE guidelines? Translate this insight to climate change.
3. When members of a group (e.g. those in your class) want to jointly organise a party, they may have very different preferences for drinks, music, etc. How do they reach

agreement? Can you transfer these insights into agreeing about climate change policies? If not, then why?

4. Are there substitutions available for all material products? What about substitutability for the natural environment or for climate?

Note

1 The label 'dismal science' comes from Carlyle (1849), while the price-value comparison comes from Wilde and Bristow (1891).

References

Arrow, K.J. and Debreu, G. (1954) Existence of an equilibrium for a competitive economy. *Econometrica*. **22**: 265–290.

Bromley, D.W. (1990) The ideology of efficiency: searching for a theory of policy analysis. *Journal of Environmental Economics and Management*. **19**: 86–107.

Carlyle, T. (1849) Occasional discourse on the negro question. *Fraser's Magazine for Town and Country* Vol. XL. London.

Cline, W.R. (1992) *The Economics of Global Warming*. Washington, DC: Institute for International Economics.

Dietz, S. and Venmans, F. (2018) *Cumulative Carbon Emissions and Economic Policy: In Search of General Principles*. Grantham Research Institute WP No. 283. London: London School of Economics.

Fraunhofer Institute for Solar Energy Systems (2018) *Photovoltaics Report*. Available at: www.ise.fraunhofer.de/de/downloads/pdf-files/aktuelles/photovoltaics-report-in-englischer-sprache.pdf

Funtowicz, S.O. and Ravetz, J.R. (1993) Science for the post-normal age. *Futures*. **25**(7): 735–755.

Gerlagh, R. and Liski, M. (2017) Consistent climate policies. *Journal of the European Economic Association*. **16**(1): 1–44.

Gollier, C. and Tirole, J. (2015) Negotiating effective institutions against climate change. *Economics of Energy and Environmental Policy*. **4**(2): 5–28.

Heal, G.M. and Millner, A. (2014) Agreeing to disagree on climate policy. *Proceedings of the National Academy of Sciences*. **111**(10): 3695–3698.

Hubin, D.C. (1989) Scarcity and the demands of justice. *Capital University Law Review*. **18**(2): 185–199.

IPCC (2014) *Climate Change 2014: Synthesis Report. Contribution of Working Groups I, II and III to the Fifth Assessment Report of the Intergovernmental Panel on Climate Change*. Geneva, Switzerland: IPCC.

Jevons, W.S. (1965) *The Theory of Political Economy*. 5th edition. New York: Augustus M. Keller.

Legal Information Institute (2017) *The Code of Federal Regulations*. Available at: www.law.cornell.edu/cfr/text/31/subtitle-B

Lo Piano, S. and Mayumi, K. (2017) Toward an integrated assessment of the performance of photovoltaic power stations for electricity generation. *Applied Energy*. **186**: 167–174.

Marshall, A. (1920) *Principles of Economics*. 8th edition. London: Macmillan.

Mayumi, K. and Giampietro, M. (2012) Response to 'Dimensions and logarithmic function in economics: a comment'. *Ecological Economics*. **75**: 12–14.

Mayumi, K. and Giampietro, M. (2018) Money as the potential cause of the tragedy of the commons. *Romanian Journal of Economic Forecasting*. **21**(2): 151–156.

Nordhaus, W.D. (1992) An optimal transition path for controlling greenhouse gases. *Science*. **258**: 1315–1319.

Nordhaus, W.D. (1993) Rolling the 'DICE': an optimal transition path for controlling greenhouse gases. *Resource and Energy Economics*. **15**(1): 27–50.

Nordhaus, W.D. (2007) Critical assumptions in the Stern review on climate change. *Science.* **317** (5835): 201–202.

Nuclear Energy Agency (2016) *Uranium 2016: Resources, Products and Demand.* OECD. NEA. No. 7301. Available at: www.oecd-nea.org/ndd/pubs/2016/7301-uranium-2016.pdf

Pearce, D. (2003) The social cost of carbon and its policy implications. *Oxford Review of Economic Policy.* **19**(3): 362–384.

Rezai, A. and Van der Ploeg, F. (2016) Intergenerational inequality aversion, growth, and the role of damages: Occam's rule for the global carbon tax. *Journal of the Association of Environmental and Resource Economists.* **3**(2): 493–522.

Samuelson, P.A. and Nordhaus, W.D. (2010) *Economics.* 19th edition, New York: MacGraw-Hill.

Stern, N.H. (2007) *The Economics of Climate Change: The Stern Review.* Cambridge: Cambridge University Press.

Stiglitz, J.E. (1997) Reply: Georgescu-Roegen versus Solow/Stiglitz. *Ecological Economics.* **22**: 269–270.

Tainter, J.A., Taylor, T.G., Brain, R.G. and Lobo, J. (2015) Sustainability. In R. Scott and S. Kosslyn, eds. *Emerging Trends in the Social and Behavioral Sciences.* Hoboken, NJ: John Wiley & Sons.

van Den Bijgaart, I., Gerlagh, R. and Liski, M. (2016). A simple formula for the social cost of carbon. *Journal of Environmental Economics and Management.* **77**: 75–94.

Vitousek, P.M., Ehrlich, P.R., Ehrlich, A.H. and Matson, P.A. (1986) Human appropriation of the products of photosynthesis. *BioScience.* **36**(6): 368–373.

Wilde, O. and Bristow, J. (1891) *The Picture of Dorian Gray.* 2006 edition, Oxford: Oxford University Press.

Part II
What should we do?

6 Are carbon markets the best way to address climate change?

Misato Sato, Timothy Laing and Mike Hulme

Summary of the debate

Using market-based mechanisms to drive change in energy investment and individual behaviours has been widely advocated and increasingly implemented over the last 20 years. **Misato Sato** and **Timothy Laing** argue in favour of carbon markets because they embed carbon prices in the realities of economic and political systems. Carbon markets are spreading rapidly into new regions and sectors and these authors defend the view that they will greatly advance the reduction of future carbon emissions. **Mike Hulme** challenges this position by exposing the limitations of neoclassical economic orthodoxy as a justification for emissions trading and by offering evidence of the failures of carbon markets in practice. Furthermore, carbon markets do little to bring about deeper structural or behavioural changes which are necessary for adequately mitigating climate change.

YES: Markets are flexible, efficient and politically feasible (*Misato Sato and Timothy Laing*)

Introduction

From an economic perspective climate change is essentially an externality problem. Global economic growth to date has been (by and large) fuelled by fossil fuels and greenhouse gas (GHG) emissions that arise from almost all economic activities, whether power generation, transport, agriculture, industry or retail. Such growth has brought prosperity, particularly in the Global North, but increasingly too in other parts of the world. At the same time it has created an externality problem in the form of global climate change, the damages from which will be felt largely by others, particularly future generations. In other words, if there is no carbon pricing in the economy and if we fail to consider the full costs of carbon emissions in our decision-making, those of us who make use of the finite carbon resource will not have to pay the consequences of it.

Traditional economic theory suggests that the solution to such problems is to price the externality and ensure that those who cause the problem pay for it. If priced correctly to reflect the environmental damage (see **Chapter 5**), then a carbon price will shift behaviour and markets will deliver an efficient, optimal outcome. In practice, carbon prices are a necessary, but not sufficient, solution. The process of adjusting to higher prices can be slow, painful and complex, such that carbon pricing needs to be complemented with other measures, including standards and strategic investments. Yet carbon pricing is likely to play a central role in mitigating climate change, especially in market-based economies.

Carbon can be priced through taxation or by issuing a fixed quantity of emissions allowances that can be traded (terms used for the latter include 'carbon markets', 'cap-and-trade' and 'emissions trading systems' (ETS) and these will be used interchangeably in this essay). The merits and demerits of both options have been extensively discussed in the economics literature (see Goulder and Schein, 2013, for a comprehensive review). With a carbon tax, the government sets the price and collects the revenue. Under cap-and-trade, however, the government sets the emissions cap, i.e., the target level of emissions. It then issues a corresponding number of emissions allowances and distributes these to participating firms, either through free allocation or auctioning. A carbon price emerges from the market, as firms sell and buy allowances depending on whether they have a surplus or shortage of allowances relative to their emissions. The market price encourages all firms to seek out and take up low cost mitigation opportunities, even if they receive the allowances for free. This is because if firms find abatement options that costs less than the market price for carbon, they can sell unused allowances on the market for a profit. On the flip side, if no low cost abatement opportunities are available, then firms can purchase allowances from the market to keep emitting. In this way, the overall emissions cap is met at least cost to society.

Emissions trading systems (ETS) have made great headway. In 2018, there were 20 ETSs worldwide, operating at different scales, including regional, national and subnational. They covered about 8% of global emissions, while carbon pricing as a whole covered about 14% (World Bank and Ecofys, 2018). The largest ETS to date is the European Union's Emissions Trading System (EU ETS), which has been operating since 2005, although this will soon be surpassed by China's national ETS (Wang et al., 2019).

Why carbon markets are effective

In what follows we promote six reasons—a mixture of both political and economic— why carbon markets are a good way for addressing climate change through the reduction of carbon emissions.

Prices matter

The most compelling argument for using carbon markets, and carbon pricing in general, is that prices matter. When prices convey the true cost of a good or service, including the cost of environmental damages entailed, then we can make an informed decision about whether we are willing to buy that good or service. A carbon price changes relative prices, making carbon intensive activities more expensive and driving people away from carbon intensive purchases towards low carbon alternatives. Markets will then respond by innovating and delivering new lower carbon technologies. The case for using price instruments is particularly compelling for carbon because emissions originate from a broad range of activities throughout the economy and have innumerable options for control. For governments to pick technologies or to micro-manage all opportunities is intuitively infeasible and undesirable (see **Chapter 7**).

A key advantage with carbon pricing is that it is technology neutral and aids the discovery of new processes. The experience in the USA in the 1990s with regulating sulphur from coal power generation (the acid rain problem) through the sulphur dioxide cap-and-trade scheme demonstrated this. Despite claims from industry that technology options were limited and expensive, pricing sulphur dioxide emissions created a profit-motive to not only adopt readily available technologies, but also to find new solutions to the problem. Through a process of market discovery, cheap technological and non-technological solutions were found, including innovation in production processes, organisational behaviour, as well as sourcing of low-sulphur coal (Taylor et al., 2005).

Indeed, there is powerful evidence that people and businesses do respond to higher energy prices and that economies adjust to these prices over time (Grubb et al., 2018). Countries with relatively high energy prices pay more in the short-term, but over time they adapt by using energy more efficiently to produce economic output. These gains from using energy efficiently can be so large that countries with high prices end up spending less overall on energy compared to countries with low prices. As carbon pricing essentially works by increasing energy prices, it also helps accelerate the development of low carbon energy sources and the shift towards the production and consumption of goods and services with low energy and carbon footprints.

Markets are more politically expedient than a carbon tax

One of the main advantages of carbon markets over carbon taxation is that they have proved more politically expedient. While there have been some examples of implementing carbon taxes, many attempts to enact such carbon prices encountered major political resistance and ended in failure. In Europe, a succession of proposals were made to implement an EU-wide harmonised carbon tax in the early 1990s, but these failed despite the various concessions made to opposing industry groups and individual member states. In the USA, the Obama administration (2008–2016) tried arguing that an economy-wide carbon tax was better than the Clean Power Plan, but it didn't go far. In the USA, as in many other countries, 'tax' is a dirty word, rendering a carbon tax a hard-sell. Where carbon tax proposals have survived, for example in Norway, Sweden, Finland and Denmark, they are usually watered down and softened with exemptions.

Cap-and-trade on the other hand has proved more palatable as a regulatory instrument across many contexts. In addition to its non-tax status, a number of key features of carbon markets make them attractive to policymakers. In particular, carbon pricing is usually met with strong opposition from industry sectors who argue that the regulatory cost burden will damage their international competitiveness. Carbon markets, however, offer a way to price carbon whilst largely shielding industry from direct and indirect costs, such as by distributing emissions permits for free, thus fostering political support (Joskow and Schmalensee, 1998). Although similar protection can be offered to industry under a carbon tax through exemptions and reduced tax rates, a system of emissions trading with free allocation has seen more success in tempering political opposition from highly mobilised producer groups. This is perhaps because firms entitled to free allocation can not only avoid high regulatory cost burden, but also enjoy revenues by passing costs on to customers whilst receiving allowances for free, making profits from the difference.

Another problem with carbon taxation is that the difficult task of setting and adapting the tax rate appropriately over time rests with the government. With cap-and-trade, the price setting is instead left to the market. This separation frees governments from having to implement unpopular tax raises—as was attempted by President Macron in France in

2018—and also reduces the risk of sudden policy changes for companies when governments change. With regional carbon pricing initiatives such as in Europe, carbon markets provided a solution when individual governments were unwilling to hand over control of setting tax rates to another legislative body, i.e. the European Union. Indeed, whereas an EU-wide carbon tax is considered a fiscal matter requiring unanimity between all EU member states, emissions trading falls within the realm of environmental policy, thus requiring only a qualified majority. This eases the political process for instigating markets.

Markets offer more certainty over environmental outcome

From an environmental perspective, a key advantage of carbon markets is the certainty of the outcome. As implied by the term cap-and-trade, carbon markets effectively set a limit on emissions. This 'cap' is set by policymakers and can gradually decline with time to align with commitments countries have made nationally or under international climate agreements. With policy instruments such as carbon taxes, emissions standards or innovation funding, the environmental outcome is less certain.

Price uncertainty can be managed with price control mechanisms

A corollary of certainty over quantity outcomes is the uncertainty over price. Under cap-and-trade, carbon prices and hence the overall cost of mitigation can be less certain than under a carbon tax. Long-term uncertainty over carbon prices can erode investor confidence and be problematic for driving forward large-scale investments into low carbon technology and assets. However, carbon markets can provide price stability and manage investment risk by embedding price control mechanisms into the carbon market design. For example, California's cap-and-trade system has a gradually rising price floor (below which prices can't fall) and a price ceiling (above which they won't rise). This creates a 'price collar', stabilising costs for industry. Large price fluctuations can also be avoided by allowing banking of permits across time (Schmalensee and Stavins, 2017) or through mechanisms that absorb surplus allowances such as the market stability reserve in the EU ETS.

Greater efficiency in the long-term

An economic case can also be made for carbon markets vis-à-vis a carbon tax. Under strict idealised circumstances, taxes and trading will deliver the same outcome in terms of emissions. But in reality, both markets and information are imperfect. Particularly when it comes to a complex and long-term problem like climate change, the associated uncertainties are huge. This makes it difficult to know how much environmental and economic damage will occur with temperature rise (see **Chapter 5**) and how much it will cost to keep emissions down.

On the science side, predicting how much environmental damage will result from a certain level of emissions is highly uncertain because there may be tipping points in the climate system (see **Chapter 2**). This means that in the long run, there may be some stock level beyond which damages increase steeply with emissions. On the economics side, predicting the costs of emissions reductions is difficult especially in the short-term, because of the inherent uncertainties with technological development. It is hard to know which technologies will 'win' and deliver the necessary emissions reductions, and how much they will cost, because some technology solutions are yet to be invented and proven.

Weitzman (1974) showed theoretically that under conditions of uncertainty, taxes are more efficient in the short-term when uncertainty around costs is greater. In the long run, however, the uncertainty around costs falls because visibility over the technological pathway improves with time, but the uncertainty around environmental damages increases. As carbon markets ensure emissions are capped, in the long run they are more favoured than a tax.

Carbon markets are pliable, giving them greater chance of success

Another key feature of emissions trading is its pliability. Achieving a carbon tax that is well designed enough to be politically acceptable seems possible on relatively small scales, either in small and homogeneous geographical regions or for specific sectors within a political jurisdiction. In contrast, the global experience to date has proved that carbon markets can be plausibly implemented in all shapes and sizes, big and small. They can range from small city-scale schemes covering the commercial and residential buildings sector, such as the Tokyo ETS, to large regional trading schemes like the EU ETS which covers electricity generation and energy intensive industries. This flexibility arises because the variables that go into designing carbon markets—such as how to allocate permits, the stringency of the cap, the use of auction revenue and the level of minimum price—make it possible to fine tune the market to enhance political acceptability and adapt to problems as they arise.

Indeed, the European experience shows us that, even with high bureaucratic capacity and highly integrated political and financial institutions, it is unlikely that the design of a carbon market will be right from the outset. Each successive phase of the EU ETS, from 2005 onwards, has revealed new problems, including windfall profits, surplus allowances, persistently low prices and so on. But the European experience has also showed that lessons can be learned and incorporated into the design to strengthen the functioning and delivery of carbon markets over time (Newell et al., 2013). Learning-by-doing should be built into the design architecture of carbon markets to give them a greater chance of success by giving them opportunities for periodic review and reform.

Carbon markets are also conducive to global cooperation and coordination on climate change action. For example, the EU ETS allows the partial use of international credits from mitigation actions undertaken in developing countries. Carbon credits generated through both the Clean Development Mechanism and Joint Implementation projects can be used for compliance until 2020. Different carbon markets can also be linked to scale-up efforts and widen mitigation options to further increase efficiency gains, for example as has been done by several schemes in North America.

Additionally, carbon markets can be tied in with other policies necessary to accelerate the low carbon transition. For example, permit auction revenues can be used for public investments, such as innovation support for the development of new low carbon technologies (see **Chapter 7**). Revenues can also be used to address climate justice issues or to offset any negative macroeconomic impacts, for example by compensating individuals or groups such as the fuel poor. For example, 35% of revenues from California's cap-and-trade programme are allocated to projects targeting 'priority populations', which include disadvantaged communities, low-income communities, and low-income households (London et al., 2013).

Conclusion

A problem as large and complex as climate change has no silver bullet solutions. Carbon pricing also has its limits and a spectrum of policy measures will be necessary to move the world's energy economy away from fossil fuels. For example, a carbon price can spur innovation, but it cannot transform the economy. More targeted measures, such as large public investments in energy R&D, will be needed to build the technological capacity needed to make possible a transition to a zero-carbon economy. Other policies like information provision and emissions standards are likely needed to nudge businesses to make smarter choices. Nonetheless, we are convinced that carbon pricing is likely to play a central role in addressing climate change. Carbon markets are a key vehicle for signalling the real cost associated with emitting carbon and for internalising the externality for millions of individual decision-makers.

Pricing carbon through emissions markets has proved far from easy, but it is making more rapid headway than carbon taxation. It offers a solution that can be tailored to size, scale and scope of particular local circumstances. It also offers greater certainty on the key dimension of emissions reductions, more so than the alternative tool of carbon taxes. Carbon markets may not seem the most elegant solution on paper, but it seems to be the one that most likely survives the political process and hence is most implementable in the real world.

Carbon markets are now proliferating across different political contexts (World Bank and Ecofys, 2018), with the key advantage that they can induce relative price changes in the economy while shielding industry from revenue transfers and alleviating industry objections. Carbon markets can also link across national borders and also with other policies. Although they are not a sufficient solution by themselves, they are a necessary part of any policy mix to address the problem of climate change. Carbon markets help internalise the externality of GHG emissions and offer political and economic flexibility, along with other incentives, to develop and deploy the necessary energy technologies to address the climate change challenge.

NO: Carbon markets are theoretically flawed and practically ineffective (*Mike Hulme*)

Introduction

The idea that carbon markets are an effective policy tool for incentivising significant reductions in carbon dioxide—and other greenhouse gas (GHG)—emissions first took root in the early 1990s. Initially promoted by economists in the United States—based on their successful experience of a similar scheme for trading sulphur dioxide emissions—carbon markets were adopted by the European Union after the failure in 1992 to agree an EU-wide carbon tax. The ideology behind carbon markets is derived from welfare economics, and the field of neoclassical economics more broadly, and was consistent with the dominant neo-liberal economics of the 1990s, namely the practices of privatisation, deregulation and marketisation. Carbon markets were therefore written into the 1997 Kyoto Protocol as the central policy tool for climate regulation and the first fully functioning large-scale carbon market commenced in 2005 with the EU Emissions Trading Scheme (EU ETS), the centrepiece of the EU's pioneering Climate Change Programme (Bailey, 2010). More recently, China's national emissions trading system was announced in December 2017 and is expected to be gradually phased in. There are now over 20 regional or national carbon trading systems worldwide.

But wherever carbon markets have been introduced they have been resisted by a wide range of civil society actors and policy think-tanks from both the political Left and the political Right (Pearse and Böhm, 2014). These critiques point out that the case for carbon markets rests on questionable definitions of market externalities (as an aberrant failure of otherwise efficient markets); on an insufficient theory of economic incentives; on a questionable approach to ecological processes; and on an inadequate view of market structures/dynamics more broadly. Carbon markets fail to incentivise the structural changes in the political energy economy that are necessary for deep decarbonisation and they perpetuate climate injustices and development inequalities. They propagate a post-political myth (Swyngedouw, 2010) that reducing carbon emissions is a technical-economic challenge that can be met without confronting the power of incumbent fossil fuel interests.

On top of these theoretical and political critiques is the mounting evidence that carbon markets just do not work; they have failed in any significant way to accelerate the reduction in carbon dioxide and other GHG emissions. Despite more than 15 years of operation, and the growing number of carbon markets in operation worldwide, global emissions of carbon dioxide continue to rise. Rather than adequately tackling climate change, the further expansion of carbon markets may instead be entrenching *obstacles* to the deep decarbonisation of the world's energy economy called for in the **Paris Agreement**. Rather than a utopian faith in the effectiveness of market mechanisms, what is needed is more direct regulation of the energy economy: progressive carbon taxes, divestment from fossil fuel companies and accelerated direct investment in alternative low carbon energy technologies (see **Chapter 7**).

Why carbon markets are ineffective

My argument proceeds below by offering five reasons why carbon markets are not an effective way to deliver significant reductions in carbon emissions.

Efficiency

The emergence, salience and claimed success of carbon markets during the twenty-first century is rooted in the assumption that carbon trading provides the optimal—i.e. most cost-efficient—means to reduce greenhouse gas emissions. But the claim that market-based regulation is always more efficient than traditional regulation—e.g. Pigovian taxes—is rarely supported with empirical evidence and lacks a convincing theory of economic incentives. According to neoclassical theory, for markets to deliver on the promise of efficiency there need to be clearly assigned property rights, identical products, zero transaction costs between actors and perfect knowledge of the market. None of these conditions are at all obvious in the case of climate change and carbon emissions, as we shall see below. This is why the claim of efficiency is dubious:

> Mainstream economics recommends emissions trading schemes [ETS] on the basis that [they] can reduce abatement control costs . but the claims of efficiency gains for any regulatory instrument are far from clear or determinate. Transferring textbook predictions [into policy tools] will lead to exaggerated and unrealistic

expectations and ignore complex interactions. Claims for an ETS being the most efficient policy instrument cannot then be substantiated.

(Spash, 2010: 176)

The reality is that market efficiency criteria are frequently in conflict with the structural power of fossil fuel capital, concerns about economic competitiveness and with human development goals. The latter has most powerfully been seen in the case of REDD+, a market flexibility mechanism (see below) intended to reduce deforestation and degradation in tropical countries. Taken together, these concessions reflect and contribute to the vulnerability of carbon markets to rent-seeking, corruption and fraudulent behaviour (see later). Carbon markets fall well short of their promise of efficiency.

Flexibility

Property rights over the atmosphere are created by governments, either through permits to emit (e.g. carbon allowances) or through rights to emissions reductions (e.g. Certified Emissions Reductions (CER)) which represent units of emissions either avoided or stored in carbon 'sinks'. Once issued, these 'rights' are then put up for sale with the intention being that the market then decides on the trading price. One of the central justifications for a market over a tax is that emissions trading delivers a fixed 'cap' on emissions, in effect defined by the volume of government backed permits issued. But this claim is undone by the widespread use of carbon offsets and permit banking in the design of carbon markets. (*Carbon offsetting* allows companies to claim the benefit of hypothecated future emissions reductions they have 'secured' in geographical distant locations; *permit banking* allows companies to roll-over unused allowances into later phases of the market scheme.)

These so-called 'flexibility mechanisms' were introduced into the Kyoto Protocol in order to expand the range of venues, actors and processes that could participate in the market (and hence deliver on the promise of least-cost). They also opened a politically expedient way for financial transfers from richer nations to developing countries, notably through the Clean Development Mechanism (CDM) and Joint Implementation (JI) projects. But there are problems with such flexibility mechanisms. These include the spatial displacement (so-called 'leakage') and temporal displacement of emissions reduction (permit banking), the reliance on counterfactual scenarios to establish what is a valid offset and heroic and unprovable assumptions about the long-term integrity of carbon sinks.

All of these facets of market flexibility give cause to question whether carbon markets are delivering real or merely nominal reductions in emissions. For example, Branger et al. (2015) cite the example of the recycling of 800,000 CERs by the Hungarian Government, legal within the terms of the CDM but in effect resulting in the double-counting of emissions reductions and Lohmann (2009) points out the more fundamental impossibility of establishing the 'additionality' of CDM or JI projects. The possibility of fraud and corruption are built into the design of such market-based mechanisms. It is with good reason that carbon offsetting through the CDM has been roundly criticised as 'a loophole and an unjust form of mitigation that distracts from the central task of transitioning away from fossil fuel extraction in the North' (Pearse and Böhm, 2014: 329).

States and markets

A third argument against the effectiveness of carbon markets concerns the nature of the relationships between capital, fossil fuel incumbency and the state that are embedded in their design. Rather than allowing the free operation of buyers and sellers in a neutral market, the design of carbon markets is significantly the result of lobbying by powerful vested interests and industries (see **Chapter 12**). Emissions trading schemes are distorted by conditions of near-monopoly and by the concentrated power of fossil fuel companies. This leads to an embellished faith in manipulated price signals, which serve as justification to avoid more substantial economic reform. It is ironic that a 'free-market' economy such as the United States has rejected cap-and-trade, whilst the command-and-control economy of China has embraced carbon markets (see **Chapter 14**). In Europe the political and economic power of fossil fuel companies and the Energy Intensive Trade Exposed (EITE) business sector has diluted the effectiveness of the European carbon market. Numerous concessions have been offered.

Notable among these has been the free and generous allocation of emissions allowances. Although economic theory is clear that auctioning is the best way to distribute such permits, such a process has high initial implementation costs and would meet significant political resistance. In practice, only a very small proportion of allowances in any ETS are auctioned. Further concerns relate to the issues of 'carbon leakage' and industrial competitiveness (Branger and Quirion, 2014). *Carbon leakage* is the notion that differential carbon prices across borders and between regions lead to the production of carbon intensive goods moving from countries within a carbon market to so-called 'carbon havens', countries with no such carbon constraints. Branger et al. (2015) estimate a carbon leakage ratio of between 5 and 25% of the 'controlled' emissions in the EU ETS.

Taken together, the above concerns have meant that in most ETS schemes—including those in the EU and China—rather than auctioning allowances, governments allocate them according to the historical emissions of the participating companies, called 'grandfathering'. European companies participating in the EU ETS have benefitted from significant windfall profits through the over-allocation of permits resulting from this process, whilst at the same time passing on the costs of market participation to consumers. For example, during the first two phases of the EU ETS—2005 to 2012—energy intensive industries in Europe enjoyed windfall profits from the over-generous allocation of allowances of around €35 billion (Branger et al., 2015; Pearse and Böhm, 2014).

A further reason for the poor performance of carbon markets concerns price volatility. States have found it impossible to institute reliable—and high—carbon prices that are capable of providing incentives for meaningful change to the structure of energy production. For example, the price of carbon in the EU ETS has varied from nearly €40 per tonne to as low as €3 per tonne. The result of this volatility has been 'price-fixing' by market regulators using the devices of floor, ceiling and collar prices for carbon in order to offer some semblance of stability for traders (Schmalensee and Stavins, 2017). The newly emerging national emissions trading scheme in China is especially vulnerable to such volatility 'due to the country's economic structural transition from a manufacturing-based to a service-based economy' (Wang et al., 2019: 4). If carbon markets are to operate successfully in a command-and-control political economy such as China (see **Chapter 14**), then significant reforms and institutional strengthening will be necessary. China will also need much more credible and robust systems of emissions measuring, reporting and verification (MRV), since no market can function without believable commodity and compliance data.

Distraction

Emissions trading by design privileges only least-cost emissions reductions, not structural transformation. It therefore fails to adequately incentivise long-term planning by the relevant industry actors for the future energy transition away from carbon-based fuels. Rather than being technology neutral, carbon markets are instead in danger of perpetuating carbon-based energy systems by placing obstacles in the way of deep decarbonisation. Rather than being a first step toward broader, deeper and cleaner technology reform, carbon trading 'locks in emissions increases and is used as an excuse to abandon other energy policies that contribute more meaningfully to the task of decarbonisation' (Pearse and Böhm, 2014: 333).

In his consideration of the ethics of emissions trading, Edward Page (2013) regards this inevitable, if unintentional, consequence of carbon markets as *unethical* because it erodes the intrinsic motivations of emissions actors to protect the climate system. This 'crowding-out' effect of emissions trading ends up being perverse:

> The diffusion of [this effect] amongst participants and nonparticipants … will eventually prove counter-productive to the health of the environment, since it will hinder the development of an ethos of environmental concern by conditioning agents to expect extrinsic rewards to protect natural objects and systems that they would ordinarily be motivated to protect for non-monetary reasons.
>
> (Page, 2013: 241)

Carbon markets, under the guise of dealing with climate change 'efficiently', in fact end up promoting a form of 'moral corruption', to use Stephen Gardiner's memorable term (Gardiner, 2011).

Politics

Carbon markets are premised on the neo-liberal logic of the commodification of nature and on the neoclassical economic assumption of optimal allocation of scarce resources (see **Chapter 5**). Abstract value is assigned to the atmosphere through the use of marginal abatement costs, i.e. the cost of reducing one more unit of emissions. Carbon markets are therefore constructed and operated by complex socio-technical networks of economic agents, monitoring technologies and accounting systems, creating a carbon technocracy which is remote from democratic politics and unaccountable to it. These new managerial elites are invested in the propagation of carbon markets and in defending public faith in technocratic expertise to deliver public benefits through the markets (Lohmann, 2009). The complexity of carbon market governance escapes scrutiny from elected politicians, even more so from the people, the demos. This explains the widely recognised extent of corruption and fraud within the first decade of the EU ETS (Lohmann, 2010). For example, VAT frauds in the EU ETS in 2009/10 alone resulted in €5 billion of lost tax revenues. 'The relative immaturity of the market and the intangible nature of allowances make carbon markets particularly vulnerable [to fraud] compared to other markets' (Branger et al., 2015: 12).

For these and other reasons, the technocratic nature of carbon trading has met resistance from both the political Left and the political Right for being, respectively, either too unregulated or too regulated. The Left complains about the post-political nature of carbon markets—their lack of transparency and accountability—and the way in which they commodify nature, protect incumbent fossil fuel interests and encourage fraudulent behaviour. Carbon markets are frequently regressive in fiscal terms—a trigger for the political Left—in that the burden of extra carbon costs are disproportionately borne by low-income households.

The Right resists carbon markets—as we have seen in the USA and in Australia (Crowley, 2017)—because of their *over*-regulated character which interferes with the ideology of unfettered market operation. The American Clean Energy and Security Act of 2009 which proposed a cap-and-trade system for the USA—otherwise known as the Waxman-Markey Bill—was never brought to the Senate for approval in the face of stiff opposition from the Republican Right. Far from being politically expedient, these examples show that, at least in some jurisdictions, carbon markets have contributed to policy deadlock.

Conclusion

Relying on carbon markets to drive forward significant reductions in carbon emissions places an unwarranted degree of faith in the ideology of neoclassical economics which underpins such market-based mechanisms. The assignment of property rights is subject to intense lobbying from rent-seeking actors, not all industry actors are optimisers, not all greenhouse gases are commensurable and marginal abatement costs are poorly known. The marketisation of climate policy through carbon markets has developed as a strategy for states to manage the climate contradiction of capitalism. In essence, carbon markets perpetuate the systematic destruction of ecosystems that capitalism has relied upon.

Over nearly two decades of practice, numerous experiments in emissions trading have led to very little substantive change in the emissions profiles of industrial sectors and nations where ETS and offsets operate. Globally, they have achieved next to nothing. And worse, carbon markets have worked *against* the prospect of a popular, fair and effective transition away from fossil fuels. As Pearse (2017: chapter 8) argues, 'The institution of new carbon markets creates new arenas of conflict, for instance over the regressive impacts of carbon pricing on the national economy, polluter windfall profits and the appropriation of carbon rents in the South'. Environmental mobilisations beyond the debate about carbon prices opens up alternative possibilities for progressive climate politics (see **Chapter 15**). Rather than a utopian faith in the effectiveness of market mechanisms, what is needed is more direct regulation of the energy economy: progressive carbon taxes, divestment from fossil fuel companies and accelerated investment in low carbon alternative technologies. An emerging energy justice agenda and alternative strategies for contesting the state's role in energy market transformations holds promise.

Acknowledgements

The argument presented here is based on brief notes provided by Dr Rebecca Pearse, University of Sydney, who was the intended author of this essay, but was unable to complete the assignment in time for the book's publication.

Further reading

Ellerman, A.D., Convery, F.J. and de Perthuis, C. (2010) *Pricing Carbon: The European Union Emissions Trading Scheme*. Cambridge: Cambridge University Press.

This book provides the best book-length treatment of the world's first significant carbon market, the European Union's Emissions Trading Scheme. The authors explain why the scheme generated so much controversy, especially during its early years, and allows you to get behind the headlines and gain an understanding of how and why the market was constructed the way it was.

Grubb, M. (2014). *Planetary Economics: Energy, Climate Change and the Three Domains of Sustainable Development*. London: Routledge.

This book provides a comprehensive analysis of the global challenges of energy, environment and economic development. The radical changes needed in the energy system involve action in three distinct domains: markets and pricing; standards and engagement; and strategic investment. All three are necessary, but not individually sufficient. The book thus highlights the importance, but also the limitations, of markets as a tool.

Page, E.A. (2013) The ethics of emissions trading. *WIREs Climate Change*. **4**(4): 233–243.

This article offers a helpful critical introduction to the ethical issues raised by the idea of carbon markets and emissions trading. Four areas of ethical reasoning are explored where emissions trading schemes have been considered vulnerable to critique and which go beyond the dominant normative desiderata of environmental efficiency and cost efficiency.

Michaelowa, A. and Shishlov, I. (2019) Evolution of international carbon markets: lessons for the Paris Agreement. *WIREs Climate Change*. **10**(6): e613.

This review article offers a nice historical summary of the emergence and evolution of carbon markets, from the 1990s to the present day, and argues that the Paris Agreement will greatly benefit from past experience with international carbon market mechanisms. The authors identify four phases in this evolution: 'emergence', 'the gold rush', 'fragmentation' and 'post-Paris'.

Pearse, R. (2017) *Pricing Carbon in Australia: Contestation, the State and Market Failure*. Abingdon: Routledge.

Pearse argues here that the experiment in carbon trading is failing because carbon market schemes have been plagued by problems and resistance from both the political Left and Right. Using the example of Australia, and other international comparisons, the book offers a critique of the political economy of marketised climate policy, exploring why she believes the hopes for global carbon trading have been dashed.

Stavins, R.N. (2008) A meaningful U.S. cap-and-trade system to address climate change. *Harvard Environmental Law Review*. **32**(2): 293–371.

Robert Stavins—one of the original proponents of carbon markets—provides a clear description of the abortive carbon trading system for the USA. Nevertheless, it offers a useful guide to the design of an economy-wide CO_2 cap-and-trade system and is compared with frequently discussed alternatives. Common objections to a cap-and-trade approach to emissions reductions are outlined and responses offered.

Follow-up questions for use in student classes

1. Do you think that carbon markets offer an effective democratic policy tool for tackling climate change? Why do they attract criticism from both the political Left and the political Right?
2. Under which theoretical assumptions do carbon markets and carbon taxes deliver the same outcome? In which ways are these assumptions violated in the real world?

3. What do you think are the merits and demerits of carbon markets vis-à-vis carbon taxes from: (a) the perspective of regulated firms: and (b) the perspective of the government policymaker?

4. Beyond carbon markets, what other types of policies do you think are necessary to accelerate the transition towards a low carbon economy? Why are these necessary?

References

Bailey, I. (2010) The EU Emissions Trading Scheme. *WIREs Climate Change*. **1**(1): 144–153.

Branger, F., Lecuyer, O. and Quirion, P. (2015) The European Union Emissions Trading Scheme: should we throw the flagship out with the bathwater? *WIREs Climate Change*. **6**(1): 9–16.

Branger, F. and Quirion, P. (2014) Climate policy and the 'carbon haven' effect. *WIREs Climate Change*. **5**(1): 53–71.

Crowley, K. (2017) Up and down with climate politics 2013–2016: the repeal of carbon pricing in Australia. *WIREs Climate Change*. **8**(3): e458.

Gardiner, S.M. (2011) *A Perfect Moral Storm: The Ethical Tragedy of Climate Change*. Oxford: Oxford University Press.

Goulder, L.H. and Schein, A.R. (2013) Carbon taxes versus cap and trade: a critical review. *Climate Change Economics*. **4**(3): 1350010.

Grubb, M., Bashmakov, I., Drummond, P., Myshak, A., Hughes, N., Biancardi, A., Agnolucci, P. and Lowe, R. (2018) *An Exploration of Energy Cost, Ranges, Limits and Adjustment Process*. Technical Report, Institute for Sustainable Resources. London: University College London.

Joskow, P.L. and Schmalensee, R. (1998) The political economy of market-based environmental policy: the US Acid Rain Program. *Journal of Law and Economics*. **41**(1): 37–83.

Lohmann, L. (2009) Regulation as corruption in the carbon offset markets. In S. Böhm and S. Dabhi. eds. *Upsetting the Offset: The Political Economy of Carbon Markets*. London: Mayfly Books. pp. 175–191.

Lohmann, L. (2010) Uncertainty markets and carbon markets: variations on Polanyian themes. *New Political Economy*. **15**(2): 225–254.

London, J., Karner, A., Sze, J., Rowan, D., Gambirazzio, G. and Niemeier, D. (2013) Racing climate change: collaboration and conflict in California's global climate change policy arena. *Global Environmental Change*. **23**(4): 791–799.

Newell, R.G., Pizer, W.A. and Raimi, D. (2013) Carbon markets 15 years after Kyoto: lessons learned, new challenges. *Journal of Economic Perspectives*. **27**(1): 123–146.

Page, E.A. (2013) The ethics of emissions trading. *WIREs Climate Change*. **4**(4): 233–243.

Pearse, R. (2017) *Pricing Carbon in Australia: Contestation, the State and Market Failure*. Abingdon: Routledge.

Pearse, R. and Böhm, S. (2014) Ten reasons why carbon markets will not bring about radical emissions reduction. *Carbon Management*. **5**(4): 325–337.

Schmalensee, R. and Stavins, R.N. (2017) The design of environmental markets: what have we learned from experience with cap and trade? *Oxford Review of Economic Policy*. **33**(4): 572–588.

Spash, C.L. (2010) The brave new world of carbon trading. *New Political Economy*. **15**(2): 169–195.

Swyngedouw, E. (2010) Apocalypse forever? Post-political populism and the spectre of climate change. *Theory, Culture and Society*. **27**(2–3): 213–232.

Taylor, M.R., Rubin, E.S. and Hounshell, D.A. (2005) Control of SO_2 emissions from power plants: a case of induced technological innovation in the U.S. *Social Change*. **72**: 697–718.

Wang, P., Liu, L., Tan, X. and Liu, Z. (2019) Key challenges for China's carbon emissions trading program. *WIREs Climate Change*. **10**(5): e599.

Weitzman, M.L. (1974) Prices vs. quantities. *The Review of Economic Studies*. **41**(4): 477–491.

World Bank and Ecofys (2018) *State and Trends of Carbon Pricing 2018*. World Bank Group Report.

7 Should future investments in energy technology be limited exclusively to renewables?

Jennie C. Stephens and Gregory Nemet

Summary of the debate

This debate opens up questions about energy transitions and future investment in technology innovation. **Jennie Stephens** argues that it is entirely feasible to meet future energy demands using existing or foreseeable renewable technologies, transitioning the world's energy economy away from fossil fuels without needing to rely on nuclear or carbon capture and storage. Energy investments should be exclusively limited to renewable energy technologies; anything else is a distraction. **Gregory Nemet** challenges this view as hypothetical at best, risky at worst. He argues that a broader energy investment portfolio—including in nuclear fission and carbon capture and storage—is a more robust strategy for delivering the transition away from fossil fuels and will actually allow societies to better grasp the opportunities that renewable energies provide.

YES: Accelerating energy transformation requires a commitment to ending fossil fuel investments (*Jennie C. Stephens*)

Introduction

The societal transformation to renewable based energy systems is well underway, but this transition needs to be accelerated to respond to the growing threats to humanity associated with climate change. The IPCC's 2018 Special Report on 1.5°C emphasised both this urgency and the massive scale of energy system change that is required (IPCC, 2018). At this point, a necessary component of accelerating the renewable revolution is ensuring that all future investments in energy technologies are limited exclusively to renewables. Limiting energy investments to renewable technologies, which include grid technologies and storage technologies that facilitate renewable based energy systems, is crucial. Investing in other types of energy technologies, particularly fossil fuel infrastructure, will slow down the transition by perpetuating fossil fuel reliance.

Given both the urgency and the scale of energy system change that is required, a deep and consistent societal commitment must be made to a future based on renewable energy. Renewable energy (RE) includes a diversity of different energy sources that can be integrated to meet evolving future demand of different regions of the world. A renewable based future includes solar, wind, hydro, geothermal, tidal and wave technology deployed at multiple scales (large-scale, small-scale and mid-scale). Different regions of the world, and different communities within any region, can deploy a regionally appropriate mix of renewable energy technologies that harnesses the renewable resources of that region to satisfy demand. The diversity of renewable energy options, each of which can be modularised at different scales, allows for a locally appropriate heterogeneous mix of distributed energy sources.

Future renewable based systems do not have to replicate the spatial distribution of the legacy centralised systems that have dominated energy systems in the past. These new energy systems can be locally and regionally managed empowering communities and organisations to be self-sustaining (Burke and Stephens, 2018). The flexibility and diversity of renewable energy means that the renewable transformation provides an opportunity for redistributing economic opportunities, jobs and political empowerment in ways that are more equitable (Stephens, 2019). Implementing this vision of a more healthy and resilient renewable based future requires that all future energy investments are aligned with, and exclusively focused on, advancing this vision.

The question of how to manage future energy investments is controversial because many powerful individuals and institutions who are benefitting from the legacy fossil fuel system are strategically resisting change toward renewables (Supran and Oreskes, 2017). Many are not prepared to support the disruptive social change that is possible if investments are limited exclusively to advancing a renewable based energy future (REN21, 2017). The scale and scope of this change is bigger than many are willing to accept and resistance includes making strong claims based on assumptions learned from past legacy systems. In addition, many who are resisting a 100% renewable future fail to acknowledge the diversity of renewable technologies that are available; investing in renewable energy involves investing in a diverse portfolio of different technologies— renewable energy is not a single technology.

Given this diversity and the distributed potential of renewables, we know with high confidence that future energy systems will be very different than those used in the past (Burke and Stephens, 2018). Conventional thinking about energy investments may therefore be limiting change. Given this reality, it is essential that all energy investments are aligned with a resilient and sustainable vision for the future based on 100% renewable energy.

Five reasons why energy investments need to be limited to renewables

All future energy investments should be limited to renewables for several reasons: (1) the pace and scale of energy system change requires strategic focus; (2) investments in so-called 'clean fossil' or **carbon capture and storage** (CCS) are wasteful and provide false optimism; (3) investments in nuclear power are dangerous, uncertain and unnecessary; (4) renewable transformation provides an opportunity to redistribute political power and reduce socioeconomic and racial inequities; and (5) renewable energy is sufficient to meet future energy demand.

The pace and scale of change requires strategic focus

Deep systemic change is needed to respond to the accumulated negative global impacts of fossil fuel based legacy energy systems. Given the pace and scale of change that is needed, a strategic focus is required to meet energy and climate goals. This focus requires an intentional and consistent commitment to investing only in energy technologies that advance a renewable energy future. Investments in fossil fuel or nuclear infrastructure slow down the pace of change toward a renewable based future. Building new natural gas pipelines or new fossil fuel power plants, for example, locks in a fossil fuel future perpetuating fossil fuel reliance. Energy infrastructure lasts a long time, generally more than 50 years, so current and future energy investments have a long-term impact on energy systems of the future. The potential of a renewable based future can only be realised with a global commitment to deploying renewable technology at every opportunity in every community around the world. Without a strategic focus toward achieving this goal, the pace and scale of change will not be sufficient to ensure a sustainable and resilient future.

Investments in 'clean fossil' or carbon capture and storage are wasteful

Investments in 'clean fossil' technology, including investments in CCS, are wasteful and are slowing down the transition to a renewable future. CCS has been promoted as a 'bridging' technology to provide carbon dioxide (CO_2) reductions until non-fossil fuel energy is ramped up. But the past 15 years of substantial government investment in CCS and slow progress suggests that the challenges are many and it will take longer to build the CCS bridge than to shift to renewables (Stephens, 2014). Continued investments in CCS are being perpetuated although the optimistic promise of this technology has proven to be unsubstantiated and misaligned with political realities (Markusson et al., 2017). The amount of energy required to capture and store CO_2 is generally not adequately recognised in optimistic perceptions of the potential of CCS. This so-called 'energy penalty' has been estimated to be about 30% (with a range from 11–40%) which means roughly that for every three fossil fuel power plants utilising CCS an additional power plant would be required simply to supply the energy needed to capture and store the CO_2 (Stephens, 2014). The magnitude of this energy penalty (including even the lower estimates) is so high that it is difficult to imagine a future scenario in which consuming this much additional energy to enable CCS would actually make sense.

Optimism about the potential of CCS is based primarily on research on technical feasibility, but very little attention has been paid to the societal costs of perpetuating fossil fuels or to the sociopolitical requirements of long-term regulation of CO_2 stored underground. Continued investment in CCS and other fossil fuel technologies must end so that the distraction and complacency of the false sense of security such investments provide are removed. Instead of continuing to invest billions in the false promise of CCS, more aggressive investments should be made in technologies, policies and initiatives that will accelerate a transition to renewable based energy systems.

Investment in nuclear power is dangerous, uncertain, and unnecessary

Among some climate experts, nuclear power continues to be held up as an important future energy technology because it does not generate greenhouse gases. But nuclear power has many other negative attributes and the varied risks of continued investment in nuclear power outweigh potential benefits. It is dangerous, expensive, uncertain and unnecessary. And like fossil fuel technology, continued investment in nuclear power will slow down the transformation to a renewable based future.

Of the many risks of nuclear power, society's inability to develop an effective plan to contain nuclear waste is among the most concerning. Some energy experts explain that building more nuclear power plants right now, when there is no defined long-term plan for dealing with the nuclear waste, is like constructing a big new apartment building with no toilets. It just doesn't make sense. Nuclear power also requires uranium extraction which is limited to very specific parts of the world (see **Chapter 5**), so insecurity and geopolitical risks associated with nuclear are also prevalent.

Redistributing power and reducing inequities

The reason to limit energy investments to renewable energy is not only justified by climate change. There are many other reasons that have nothing to do with climate change to justify why all energy investments need to be limited to renewables. These other reasons include improving public health, creating jobs, limiting social and ecological destruction of fossil fuel extraction and redistributing power—both literally and figuratively—including reducing the political power of multinational fossil fuel based energy companies (Healy et al., 2019).

Limiting energy investments to renewables is also important because it provides the opportunity to enact social change including redistributing political and economic power and reducing inequities (see **Chapter 1**). Unlike fossil fuel resources or uranium resources, sources of renewable energy are abundant and widely accessible. In every part of the world, a locally appropriate mix of different renewable resources is available. This means that individuals, households and communities do not need to compete for access to a scarce resource in the same way that there is fierce constant competition for limited uncertain resources in legacy energy systems. Renewable sources of energy are, by definition, renewable and perpetual. Once initial investments have been made in technologies to harness the sun, wind, tides, water or geothermal renewable resources, individuals, households and communities can leverage a steady, low-cost 'fuel' source.

Of course the sun does not always shine and the wind does not always blow; intermittency is a challenge with renewable energy. However, a locally appropriate heterogeneous mix of renewable energy generation can overcome the challenges of intermittency by using sophisticated grid technology, energy storage and demand response that can match supply with demand. Despite intermittency, renewables are perpetual; there is a high level of certainty that the sun will travel across the sky each day, wind patterns will persist over time, water will continue to flow downhill and geothermal heat will dissipate. So although renewables require an initial investment, once that investment has been made the renewable 'fuel' (sun, wind and water) is free. Renewable energy does, of course, require some operation and maintenance, although the costs of many renewable technologies are concentrated in initial installation with minimal longer-term operation and maintenance costs.

These qualities of abundance and accessibility allow renewable energy the potential to offer long-term stability for widespread local adoption. This stability provides the foundation for future energy systems that are decentralised and much more accessible and inclusive than fossil fuel based or nuclear energy. This suggests a further justification for limiting investments to renewables. The potential for widespread renewable energy deployment around the world also facilitates a redistribution of political and economic power, jobs and a reduction in social, economic, racial and geographic inequities (Stephens, 2019). Renewable energy can be deployed in local, small-scale installations, in a dispersed and distributed way. This enables individuals, households, communities and organisations to own and manage their own energy infrastructure. This potential for distributed self-sufficiency in energy supply has transformative potential because of opportunities for local control and local economic gain. Due to the unequal global distribution of fossil fuels and uranium, non-renewable energy does not provide this same potential for transformative social change based on local and distributed ownership and control.

Renewable energy is sufficient to meet future energy demand

Exclusive investment in renewable energy is important because research shows that renewable energy could meet 100% of future energy demand (Brown et al., 2018; Jacobson et al., 2017). Those energy experts who claim that renewable energy will never be able to meet future energy demand are thinking too narrowly both about the potential of a distributed, heterogeneous mix of renewables all around the world and also about the future potential for social change including reduced expectations for energy consumption. Trends in energy consumption suggest that a future with much lower energy consumption expectations is likely. Although most efforts to model future energy scenarios assume energy demand will stay the same or increase, energy demand is actually very flexible and dependent on multiple social, cultural and technological dimensions that cannot be easily predicted. It is quite possible that energy demand could decrease significantly in the future, particularly as society co-evolves, adjusts and adapts to renewable based energy systems.

With large-scale investment in renewable energy, even dense areas of energy demand, including megacities, could be powered by renewable energy. There is an abundance of renewable energy available—the challenge is to steer sufficient investment in infrastructure to harnesses that renewable power. Investments are needed in large-scale and small-scale locally and regionally appropriate renewable technology including on-shore and off-shore wind, geothermal, solar, tidal, wave and so on. If sufficient investment is made to support widespread deployment of these technologies, we could have sufficient supply to meet demand (Jacobson et al., 2017).

There are some challenging aspects of a 100% renewable future. Envisioning such a future requires electrification of transportation. While electric trains, vehicles and buses are already mainstream in many places, electrification of shipping and aviation is more difficult to imagine. Aviation is perhaps the most challenging aspect of a renewable future. A small solar-powered aeroplane has already circumnavigated the world, so there is potential for technological advances to enable renewable based aviation. To address the aviation challenge, there are also researchers developing renewable based liquid fuels.

Conclusion

Exclusive investment in renewable energy is critically important to ensure a resilient future for humanity. Not only does the transformation away from fossil fuels (and nuclear) to a renewable based energy future reduce the risks of climate change, there are many other reasons to end fossil fuel investments. In the long run, investing exclusively in renewable energy will be cheaper than investing in a mixed portfolio. It is politically challenging right now to figure out how to finance the level of renewable energy investment that is required. But those individuals, organisations, communities, states and countries that are already investing aggressively in renewable energy are clearly going to be better off in the future because they will not be susceptible to the volatility of the costs of fossil fuels as they become more scarce. Economically, it makes sense to limit energy investments to renewables because there are so many societal costs of perpetuating fossil fuel reliance that are not adequately being incorporated into energy cost decisions (Healy et al., 2019).

Too often energy policy debates miss the critical point that renewable energy is fundamentally different to either fossil fuel or nuclear because the renewable resource (the fuel) is abundant, plentiful and perpetual. Once the initial investment is made in renewable technology to harness Earth's sun, wind and water for energy generation, competition for energy resources will be virtually eliminated. Once investments are made in harnessing renewable energy, energy can become cheap and abundant and the destructive energy geopolitics associated with fierce competition for fossil fuels will shift (IRENA, 2019). A new energy geopolitics could emerge that is based on a more equitable global distribution of energy resources.

Although advances in technology development are always possible and always uncertain, the decision of what source of fuel society should use for energy is not a question of technological innovation. It is a political, economic and social question—and there is no doubt that renewable sources of energy provide a more resilient and sustainable future than fossil fuel or nuclear energy. It is for this reason that society must move to committing that all future energy investments be limited to renewables.

NO: A diverse clean energy portfolio delivers wider social and economic benefits (*Gregory Nemet*)

Introduction

The world spends just under $2 trillion every year on energy investment (IEA, 2018), about 2% of gross world product. If we think about dealing with long-term problems like climate change, one can consider how large these investments become when accumulated over decades and growing over time. We will invest close to $100 trillion in energy by mid-century. The subject of this debate is whether all of these $100 trillion should be invested solely in renewable energy (RE). Putting all of this investment in renewables and in nothing else might work. It might allow us to decarbonise energy use affordably, bring an array of co-benefits and allow the world population to flourish and achieve modern high-energy lifestyles. But that would be a gamble and a risky one. It depends excessively on a set of conditions that we cannot assume will hold decades in the future. We can still obtain many of the tremendous benefits available from renewables by diversifying to a broader portfolio of technology pathways, which could include advanced nuclear, carbon capture and **general purpose technologies** such as batteries or artificial intelligence.

In contrast to an exclusive focus on renewables, addressing energy related challenges will require a broad array of technologies that are effective, affordable and able to be deployed on a massive scale (Iyer et al., 2015; Rogelj et al., 2015). There is no silver bullet. No one technology will do the job, not even a couple. There are several reasons to choose a broad set of technologies to address climate change and other energy related problems, rather than a small number of very promising ones. First, we don't know what will happen as renewable technologies become widely adopted and comprise high shares of energy systems. This uncertainty implies we are better off with a broad portfolio of ways to address energy problems (Anadon et al., 2017). Second, the world is a diverse place. People want different services, delivered in different ways, from their energy system (GEA, 2012). A technology that is scalable, clean and affordable in one place might not work somewhere else for a variety of reasons. Third, it is quite clear that technologies, even beneficial ones like renewables, impose adverse social impacts once deployed at very large scale (Grübler, 1998). We can avoid such negative impacts through a diverse approach to technology investment. One implication from a policy perspective is that it is undesirable to champion or concentrate on a single technological pathway. Our policies should reflect diversity as a means to support the larger social goals of energy innovation. We won't be able to do it by getting only one technology right.

Why we should spread energy investments widely

There are several reasons why spreading investment over a broad set of technologies is better for society than restricting investment to renewables alone.

Putting even half of all energy investment in renewables would be a lot

We don't need to aspire to 100% RE to obtain the benefits that renewables provide. As with many other goods and technologies, benefits diminish once technologies approach their saturation level. Getting to 80% RE, or even 50%, would be a massive change from today's 20% and would improve the world in a variety of ways (Creutzig et al., 2017). The next 20% or so after that would be far more expensive, difficult and less beneficial. Pushing RE to reach that additional 20% is inefficient and unnecessary since we have other options to reach our climate goals and meet our energy needs. We can benefit from an energy system with 'mostly' renewables, and even 'almost all' renewables, without having to deal with the difficulties, and the risks, that a pure RE system would entail.

Technological diversity is always better than concentration

A diverse mix of energy sources will be more socially beneficial than a narrow focus (Stirling, 2010). The fastest growing renewables—and those with the largest resource potential—solar photovoltaics and wind, are quite similar in their intermittence and low capacity factors. Intermittence means that RE are not always available when power is needed, rather they are available when the resource (wind or sun) is available. They are falling in cost and increasingly under-bidding fossil generation. But an energy system dominated by wind and solar would not be a diverse one. Other renewable technologies would add distinct attributes and thus contribute to the diversity of the system. For example, tidal energy is variable but predictable; geothermal energy is reliable but dispersed; and concentrating solar thermal electricity generation can store energy into the night. But these technologies are becoming relegated to niches since they have not been able to scale and generate cost reductions. Solar and wind continue to be difficult to beat. Indeed, the falling cost of batteries is increasing the value of wind and solar and diminishing the relative advantages of other renewables.

RE are also similar in that they are also characterised by low energy density, which means a large area (of land, air or sea) is needed to produce a low amount of energy. Systems tend to function better when the technologies involved are complementary. One technology's weakness may be offset by another's strength. Integrated assessment models always show higher future energy costs when the future technology mix is constrained (Bosetti et al., 2015). Important contributors to climate change mitigation include nuclear, carbon capture and bio-energy with carbon capture and storage. We add cost and add risk if we exclude these and other technologies and focus only on RE.

New problems always emerge at large scale

The global energy system is massive. And addressing the social problems this energy system creates is such an enormous undertaking that one technology—even if it is cheap, clean and reliable—is not enough. Even if it were sufficient, we would not want to rely on just one technology because of the strong benefits of diversity. At very large scales inherent to the global energy system, very large deployment always creates problems (Grübler, 1998). Solar panels can exacerbate heat islands, reduce the Earth's albedo or affect rainfall. Wind turbines can interrupt the flow of heat from the equator to the poles. Massive waste issues could emerge, as could materials requirements that limit adoption.

These and other problems can be mitigated by diversification. Work on carbon budgets makes clear that we need many decarbonisation technologies, applied to many sectors (Rogelj et al., 2018). Arguing about whether we should have 100% renewables or 80% photovoltaic (PV) misses the point. PV is at about 1% of electricity today. Growing to 50% of electricity or 30% of energy is feasible (Creutzig et al., 2017) and would be massive. But even that level leaves tremendous space for other technologies, both renewable and non-renewable.

Uncertainty over the long term demands robust approaches

Transforming energy systems, whether for climate change, national security or public health, is a long-term process that will be measured in decades. Much can change in 30–80 years. There are many unknowns in technology development, especially when deployed at the scale of the global energy system. The early enthusiasm for nuclear power provides a cautionary tale (Allen and Nemet, 2017). RE may prove similarly problematic. Other technologies we don't know about yet may emerge as more appealing. We should be willing and prepared to fund them as well.

Uncertainty is inherent in global systems, especially when measured over time horizons of decades, as will be necessary for transforming the global energy system. We know from other cases that the key to managing and making decisions under uncertainty is robustness (Baker et al., 2015). Rather than designing the optimal energy system, we need to look for solutions that work well under a wide range of possible conditions. Decisions that are robust to a wide variety of future conditions are appealing in this context. We need to keep our technological options open. More diversity creates more options and that improves the robustness of our portfolio with which to transform the energy system. Being prepared for a wide set of possible futures means that we should hedge our bets by developing other technologies beyond just RE.

Inertia means that reducing emissions is insufficient

The long lifetime of greenhouse gases in the atmosphere means that just transitioning from fossil fuels to RE will not be sufficient for achieving the stated goals of international climate agreements. Some removal of carbon from the atmosphere will be needed. RE can support carbon removal, otherwise known as **negative emissions technologies** (NETs). Indeed, it can provide the power for removal. But RE alone will not suffice to meet the temperature targets agreed upon in the **Paris Agreement**. Limiting global warming to 1.5 or even 2°C will require massive deployment of NETs (Fuss et al., 2018). Within just a few decades, NETs will need to be removing billions of tonnes of CO_2 from the atmosphere annually. There are a variety of ways to do this: planting trees, agricultural practices that retain carbon in soil, using carbon capture with bio-energy and using chemical processes to absorb CO_2 in the air. Each of these comes with costs, tradeoffs and limits. Some will require energy and that energy can and, in some cases has to be, renewable. Again, renewables can help support NETs, but limiting investment to RE limits the options we have available. NETs are a set of options we will definitely need so we cannot afford to avoid investing in them. We are going to need everything we have to address climate change. We should not try to do this with one hand tied behind our back.

We need to reduce energy demand, not just increase supply

The increasing incomes of a growing world population will lead to a rapid increase in demand for modern energy services, including transportation. RE can provide the core technology to meet that rising demand. But we also need to reduce that demand for energy—through efficiency or lifestyle change such as reducing meat consumption (Creutzig et al., 2018). Reducing our global demand for energy will make combatting our energy problems much more feasible. Those changes, whether to more efficient technology, smart cities or perhaps even lifestyle changes, will also require investment. Again, we should not limit our investment to RE.

General purpose technologies will also be needed

General purpose technologies (GPTs) are those that support the improvement and adoption of other technologies (Pearson and Foxon, 2012). Examples in the past include the steam engine, the microprocessor and the internet. The transition to a more sustainable energy system will benefit from the invention and adoption of GPTs. These might include artificial intelligence, synthetic biology or other innovations that have not yet clearly emerged as possible GPTs. For energy, we need not think of inventing new GPTs, but rather adapting them to use in a future energy system. Such adaptation will require investment. Beyond GPTs one should also be sure to consider that existing technologies, such as digitisation to enable smart cities and long-term energy storage to enable high penetration of renewables, will also require investment.

Heterogenous cultures, resources and political systems require diverse decarbonisation pathways

Not all locations are well suited to power themselves by RE alone, whether due to the supply of RE resources or due to the type of demand for electricity. For example, some places have such high density of energy demand that renewables may not be the best way to provide affordable low carbon electricity (Doherty et al., 2016). Advanced nuclear or fossil electricity with carbon capture and storage may be the most useful combination of sources for situations like these. As the number of megacities with populations above ten million grows, it will be societally beneficial to have access to technologies that can serve them quickly and efficiently. That may be RE, but it need not be.

Political economy considerations may not always favour renewables

The history of renewable energy shows that its development has been opposed repeatedly by interests whose assets would be at risk with its widespread adoption. Every major country that has seen substantial adoption of renewables has seen strong and effective action at slowing their deployment. This was the case in the USA in the 1950s and 1970s, in Japan in the 2000s, in Germany after 2010 and in China after 2011. Adoption of renewables has catalysed strong opposition from powerful coalitions of fossil energy suppliers. There is ample reason to believe that as renewables continue to grow, opposing interests will continue to act against renewables to intentionally slow their progress. Accelerating the deployment of renewables may depend on finding ways to soften that opposition rather than trying to overcome it via technological improvement and broad public support. Seriously addressing political economy considerations might thus, somewhat perversely, depend on investing in supporting those interests negatively affected by the energy transition. For example, the €200 billion subsidy programme in Germany from 2000 to 2012 that grew the solar market to today's massive scale excluded large powerful industrial energy consumers from having to pay for the subsidies. They were the ones most capable of blocking it. Instead, residential energy users paid for all of the programme.

Conclusion

It is possible to imagine a future energy system that exclusively uses RE to power human civilisation as it grows to ten billion people and beyond. Renewables are the fastest growing energy sources and have fallen in cost well below where even experts expected them to be just ten years ago. RE provides an immensely valuable resource with which to provide affordable non-polluting power without fuel price risk. RE could continue to scale up and provide co-benefits in the form of public health and economic development.

But there are many obstacles to that scenario that could prevent it from occurring. Problems will very likely emerge once RE is deployed at the scale that is necessary to power the world economy. For example, periodic resource constraints will certainly emerge as they have for all energy technologies. Land use competition could prove excessively contentious as food vs fuel competition intensifies. Storage of power over several months, rather than the several hours we have today, might turn out to be more difficult than expected. Interests with assets threatened by renewables will continue to oppose them. We need alternatives to be available that can supplement the RE system and so we need to invest in those as well. Investing all of the $100 trillion by mid-century in renewables alone is risky without devoting some proportion of that investment in complementary technologies that can hedge against adverse outcomes and that can enhance the societal and environmental value that renewables provide.

Further reading

Diesendorf, M. and Elliston, B. (2018) The feasibility of 100% renewable electricity systems: a response to critics. *Renewable and Sustainable Energy Reviews.* **93**: 318–330.

This article defends the claim that large-scale electricity systems that are 100% renewable *can* be designed to meet the key requirements of reliability, security and affordability. It also argues that transition to a fully RE system could occur more rapidly than suggested by historical energy transitions. The principal barriers to a fully RE future are claimed to be primarily political, institutional and cultural, not economic or technical.

Grübler, A. and Wilson, C., eds. (2014) *Energy Technology Innovation: Learning from Historical Successes and Failures.* Cambridge: Cambridge University Press.

This book presents a rich set of 20 case studies of energy technology innovation. It provides insights into why some innovation efforts have been more successful than others. The case studies cover a wide range of energy technologies, including successes and failures and from industrialised, emerging and developing economies.

Jenkins, K., McCauley, D., Heffron, R., Stephan, H. and Rehner, R. (2016) Energy justice: a conceptual review. *Energy Research and Social Science.* **11**: 174–182.

This article examines the question of energy justice in relation to different future energy systems. Energy justice raises questions about how the costs and benefits of energy production and consumption *should* be distributed and about whether we are being 'fair' to future generations in leaving a legacy of nuclear waste, the depletion of fossil fuels and the pollution of the atmosphere and climate.

Nemet, G.F. (2019) *How Solar Became Cheap: A Model for Low-Carbon Innovation.* Abingdon: Routledge.

Drawing on developments in the US, Japan, Germany, Australia and China, Nemet provides a comprehensive and international explanation for how solar power has become inexpensive. But he also shows how understanding the reasons for solar's success can teach us how to support other low carbon technologies with analogous properties, including small modular nuclear reactors and direct air capture of carbon dioxide.

Obama, B. (2017) The irreversible momentum of clean energy. *Science.* **355**(6321): 126–129.

Written by outgoing President Obama in 2017, this article makes the case for investing in clean energy and why the USA is well-placed to play a leading role in the necessary energy transformation. He argues that a clean energy economy is politically, economically and technologically feasible.

Follow-up questions for use in student classes

1. What are the challenges and opportunities to a city, state or country of establishing a 100% renewable goal?
2. How should fossil fuel companies engage in the renewable energy transformation?

3. What are the pros and cons of large-scale renewable energy installations versus small-scale distributed renewable energy installations?
4. What are some of the limitations of renewables? What alternatives exist?
5. Why is it that 'renewable energy' is grouped together when it involves such a diverse mix of technologies? Is it a useful category with which to make policy?

References

Allen, T. and Nemet, G.F. (2017) Energy technology exuberance: how a little humility is good for nuclear, renewables, and society [online]. *Medium* 11 July. Accessed 2 June 2019. Available at: https://medium.com/third-way/energy-technology-exuberance-how-a-little-humility-is-good-for-nuclear-renewables-and-society-f6c9ec821b2c

Anadon, L.D., Baker, E. and Bosetti, V. (2017) Integrating uncertainty into public energy research and development decisions. *Nature Energy*. **2**: 17071.

Baker, E., Bosetti, V. and Anadon, L.D. (2015) Special issue on defining robust energy R&D portfolios. *Energy Policy*. **80**: 215–218.

Bosetti, V., Marangoni, G., Borgonovo, E., Anandon, L.D., Barron, R., Mcjeon, H.C., Politis, S. and Friley, P. (2015) Sensitivity to energy technology costs: a multi-model comparison analysis. *Energy Policy*. **80**: 244–263.

Brown, T.W., Bischof-Niemz, T., Blok, K., Breyer, C., Lund, H. and Mathiesen, B.V. (2018) Response to 'Burden of proof: A comprehensive review of the feasibility of 100% renewable-electricity systems'. *Renewable and Sustainable Energy Reviews*. **92**: 834–847.

Burke, M.J. and Stephens, J.C. (2018) Political power and renewable energy futures: a critical review. *Energy Research and Social Science*. **35**: 78–93.

Creutzig, F., Agoston, P., Goldschmidt, J.C., Ludere, G., Nemet, G. and Pietzcker, R.C. (2017) The underestimated potential of solar energy to mitigate climate change. *Nature Energy*. **2**: 17140.

Creutzig, F., Roy, J., Lamb, W.F., Azevedo, I.M., De Brion, W.B. and 15 co-authors (2018) Towards demand-side solutions for mitigating climate change. *Nature Climate Change*. **8**(4): 260–263.

Doherty, M., Klima, K. and Hellmann, J. J. (2016) Climate change in the urban environment: advancing, measuring and achieving resiliency. *Environmental Science and Policy*. **66**: 310–313.

Fuss, S., Lamb, W.F., Callaghan, M.W., Hilaire, J., Creutzig, F. and 13 co-authors (2018) Negative emissions—Part 2: costs, potentials and side effects. *Environmental Research Letters*. **13**: 063002.

GEA (2012) *Global Energy Assessment: Toward a Sustainable Future* Cambridge/Laxenburg: Cambridge University Press/International Institute for Applied Systems Analysis.

Grübler, A. (1998) *Technology and Global Change* Cambridge: Cambridge University Press.

Healy, N., Stephens, J. and Malin, S. (2019) Fossil fuels are bad for your health and harmful in many ways besides climate change [online]. *The Conversation* 7 February. Accessed 2 June 2019. Available at: https://theconversation.com/fossil-fuels-are-bad-for-your-health-and-harmful-in-many-ways-besides-climate-change-107771

IEA (2018) *World Energy Investment*. Paris: International Energy Agency.

IPCC (2018) Summary for Policymakers. In: *Global Warming of 1.5°C. An IPCC Special Report on the impacts of global warming of 1.5°C above pre-industrial levels and related global greenhouse gas emission pathways, in the context of strengthening the global response to the threat of climate change, sustainable development, and efforts to eradicate poverty* [Masson-Delmotte, V., P. Zhai, H.-O. Pörtner, D. Roberts, J. Skea, P.R. Shukla, A. Pirani, W. Moufouma-Okia, C. Péan, R. Pidcock, S. Connors, J.B.R. Matthews, Y. Chen, X. Zhou, M.I. Gomis, E. Lonnoy, T. Maycock, M. Tignor and T. Waterfield (eds.)]. World Meteorological Organization, Geneva, Switzerland, 32 pp.

IRENA (2019) *A New World: The Geopolitics of the Energy Transformation*. International Renewable Energy Association Abu Dhabi. p. 93 Available from: https://irena.org/publications/2019/Jan/A-New-World-The-Geopolitics-of-the-Energy-Transformation

Iyer, G., Hultman, N., Eom, J., Mcjeon, H., Patel, P. and Clarke, L. (2015) Diffusion of low-carbon technologies and the feasibility of long-term climate targets. *Technological Forecasting and Social Change*. **90**: 103–118.

Jacobson, M.Z., Delucchi, M.A., Zach, A.F., Bauer, J., Wang, J., Weiner, E. and Yachani, A. (2017) 100% clean and renewable wind, water and sunlight all-sector energy roadmaps for 139 countries of the world. *Joule*. **1**: 108–121.

Markusson, N., Dahl Gjefsen, M., Stephens, J.C. and Tyfield, D. (2017) The political economy of technical fixes: the (mis)alignment of clean fossil and political regimes. *Energy Research and Social Science*. **23**: 1–10.

Pearson, P.J.G. and Foxon, T.J. (2012) A low carbon industrial revolution? Insights and challenges from past technological and economic transformations. *Energy Policy*. **50**: 117–127.

REN21 (2017) *Renewables Global Futures Report: Great Debates Towards 100% Renewable Energy* Paris: REN21 Secretariat.

Rogelj, J., Luderer, G., Pietzcker, R.C., Kriegler, E., Schaeffer, M., Krey, V. and Riahi, K. (2015) Energy system transformations for limiting end-of-century warming to below 1.5°C. *Nature Climate Change*. **5**(6): 519–527.

Rogelj, J., Popp, A., Calvin, K., Luderer, G., Emmerling, J., Gernaat, D. and Al, E. (2018) Scenarios towards limiting climate change below 1.5°C. *Nature Climate Change*. **8**(4): 325–332.

Stephens, J.C. (2014) Time to stop CCS investments and end government subsidies of fossil fuels. *WIREs Climate Change*. **5**(2): 169–173.

Stephens, J.C. (2019) Energy democracy: redistributing power to the people through renewable transformation. *Environment: Science and Policy for Sustainable Development*. **61**(2): 4–13.

Stirling, A. (2010) Multicriteria diversity analysis: a novel heuristic framework for appraising energy portfolios. *Energy Policy*. **38**: 1622–1634.

Supran, G. and Oreskes, N. (2017) Assessing ExxonMobil's climate change communications (1977–2014). *Environmental Research Letters*. **12**(8): 084019.

8 Is it necessary to research solar climate engineering as a possible backstop technology?

Jane C.S. Long and Rose Cairns

Summary of the debate

Solar climate engineering is a radical technological intervention proposed by some as offering a 'backstop' technology for containing the rate of future global warming. The debate revolves around the ethics and politics of researching a controversial technology. **Jane Long** argues that since we may need solar climate engineering technologies to save future lives and ecosystems we should undertake research now to become better prepared to make any future decision to deploy them. **Rose Cairns** challenges this view, arguing that continued research into solar climate engineering is a futile distraction from necessary climate mitigation. Investing in solar climate engineering research perpetuates the illusory idea that a technological fix for climate change is just around the corner.

YES: Research gives society an opportunity to act responsibly (*Jane C.S. Long*)

Introduction

People often choose either conservative or liberal types of solutions to difficult problems. However, problem-solving approaches might more usefully be divided in terms of whether people look to the past or the future. *Reactionaries* look to the past as providing a goal for the future. In this category we could put the political conservatives who deny climate change because they think the mitigation required will reduce historical freedoms and financial well-being. Consequently, they obfuscate about climate change. They are not willing to accept that we have to address climate change because their goal is to retain freedoms and wealth as they were in the past. But committed environmentalists can also be reactionary. Environmental activists originally wanted the climate focus to be solely on mitigation, which they believed would return Earth to a preferable prior state. They considered the early public discussion of adaptation to climate change as a 'moral hazard' that would distract people from mitigation. So, for many years, reactionary environmentalists obfuscated the need to adapt. They wished to manipulate societal decisions towards the one they hoped would return Earth to past conditions.

In the opposite camp, *proactionary* thinkers propose that innovation could help for new '**wicked problems**' such as climate change. Proactionary people recognise that we cannot rewind the tape and go back in time to do things differently and that we can only go forward. They think we have to face the situation we have now and prepare for the future. They certainly see that mitigation remains the most important and necessary strategy for climate change. But they recognise that the situation might evolve to incur massive damage and want to prepare strategically for unwanted eventualities. They think we should work to adapt to the changed conditions and consider the possibility of intervention to avoid the worst of the impacts, at least temporarily, in order to protect humans and ecosystems.

Like those who opposed early work on adaptation, those who oppose research on solar climate engineering fall in the reactionary category. They do not want to distract the public and policymakers from trying to return Earth to a past state. They do not trust that innovation will be in the interests of society. They don't believe that societal processes could or would make good decisions about solar climate engineering.

This essay supports proactionary research (not deployment) on solar climate engineering. **Solar climate engineering** (or solar radiation management (SRM), a form of geoengineering) could reduce some of the impacts of climate change by reflecting small amounts of inbound sunlight back out into space. This might be done in a variety of ways, including injecting reflective aerosols into the stratosphere, or spraying salt into clouds in order to brighten them and reflect more light (marine cloud brightening). SRM would not directly reduce concentrations of greenhouse gases in the atmosphere and therefore numerous expert reports have correctly concluded that it could never be a complete solution to global warming.

Importantly, SRM makes no sense without accelerated mitigation efforts, because continued emissions would require increasingly strong SRM interventions, taking the Earth further and further from any known climate conditions and correspondingly magnifying the associated risks. SRM only makes sense as a temporary measure designed to buy time for climate mitigation and for the removal of carbon dioxide (CO_2) from the atmosphere in order to avoid future catastrophes. Deployment of SRM techniques could probably lower Earth's temperature, and may save us from some climate risks, but this is not the same as returning the climate to a prior 'safe' state.

The deployment of SRM technologies could also adjust rainfall patterns, ecosystems, economic and cultural systems and agricultural productivity back towards a previous state, but the 'cancellation' would hardly be perfect. Some aspects would probably be undercompensated and others overcompensated (Ricke et al., 2010). That would create novel climates, i.e. ones not previously experienced, even in pre-industrial eras. Solar climate engineering could help reduce climate impacts for most of the planet, but some regions may not be affected positively. It will be impossible to predict the full outcomes of such intentional intervention in the climate. At best, the implementers would have to convince themselves that the risks associated with deployment were less than the risks without deployment.

So, should we learn more about what we might do with SRM technologies? I say emphatically, yes! We should use what time we have to learn as much as we can about options for the future. A decision to do research does not constitute a decision to deploy. We should choose to do research in order to know more before we might have to make a decision about whether or not to deploy. In the section below I suggest four reasons why.

More knowledge makes for better decisions

Solar climate engineering technologies may be needed to save lives and ecosystems and we should know as much as possible about them before we even contemplate any decision to deploy or not deploy at a global level. More knowledge may help to design a more effective intervention with fewer undesirable side effects, or it may show clearly that they will not be helpful or are too dangerous.

Climate change may be a disaster for life on Earth. Evidence points to impacts accumulating faster than expected. The world is not on track to stay below 1.5°C temperature increase, or even a 2°C increase (see **Chapter 2**). In a perfect world humankind would decide to implement policies to end greenhouse gas emissions as fast as possible. Successful mitigation requires major advances in energy, agriculture, transportation and industrial technology (see **Chapter 7**). A worldwide agreement and immediate action to stop all emissions in the next few decades might avoid the worst impacts. But worldwide agreement has proven elusive.

Emissions reduction (i.e. climate mitigation) at breakneck speed that avoids dangerous climate change could itself hurt many people significantly by holding back development and causing starvation (see **Chapter 1**). Clearly, we should not deliberately starve millions of people to stave off unintentional starvation of millions of people from climate impacts. As important as climate mitigation is, we have to face the fact that the world may well fall short of timely, appropriate and effective mitigation. We could definitely experience significant harm to the biosphere and humanity in the coming years. Which of course does not mean we should stop trying as hard as we can. It's just not easy to do.

Even worse, the best information we have now indicates we will have to do more than merely *stop* emissions of greenhouse gases as fast as we can to stave off disaster. For long-term climate stability we will have to actively *remove* CO_2 from the atmosphere to avoid the worst impacts of climate change. All the technologies named so far to remove CO_2 are slow and expensive, taking perhaps hundreds of years to return the climate to a previous state considered liveable (NRC, 2015). In the last round of IPCC climate model projections, every future climate simulation that managed to keep the temperature increase below 2°C did so by invoking a method—likely impractical—which would actively withdraw CO_2 from the atmosphere. Of course, invoking interventions like solar climate engineering could also exert a similar brake on global temperature increase. These results essentially show a requirement for some form of geoengineering to stay below the 2°C limit, not to mention the 1.5°C limit invoked in the recent **Paris Agreement** (IPCC, 2018).

These model predictions used by the IPCC incorporate the best scientific knowledge currently available. They do not have perfect foresight, but they do provide the best predictive information available. It seems foolish in the extreme to ignore the fact that every one of these models requires some form of geoengineering intervention to stay below putative critical temperature thresholds. The need to research a wide variety of strategies to avoid extreme climate impacts, including solar climate engineering ideas, would therefore seem clear.

Perhaps some of the best strategic thinking on the possible use of SRM was developed by John Shepherd (Long and Shepherd, 2014). In this scenario the world attempts to mitigate emissions as fast as possible, but if it were determined that something must be done to quickly reduce dangerous temperature increases, SRM would be used to temporarily achieve such an outcome. Then the use of CO_2 removal (CDR) techniques could be deployed to decrease atmospheric levels of CO_2 as rapidly as possible to the point where there is no longer a need for the SRM intervention. In this scenario, CDR could provide an exit strategy for SRM and would thus avoid the 'termination problem'—i.e. the potential for a rapid rise in temperature and resulting large climate shock if an SRM intervention were to be suddenly discontinued (Parker and Irvine, 2018). If we need to implement such a strategy and deploy SRM technologies we need to know more about them. We need research.

Learning to govern the climate responsibly

The societally responsible research of geoengineering should help society develop the same skills needed to be responsible about climate in general.

A successful proactionary solar climate engineering research agenda must be coupled with an agenda for responsible innovation that protects and honours the interests of society. We must at a minimum first learn how to govern SRM *research* if we are ever to govern a possible deployment. The first steps in research can be tightly coordinated with the first steps in governance. Developing the societal skills and techniques required to govern solar climate engineering could begin as a gradual process, through learning-by-doing rather than by trying to foresee and solve all the difficulties at the outset. SRM research governors will have to make decisions about whether or not to allow research that may itself pose some risk; the complexity of those decisions will evolve over time. At first, they will only have to deal with the relatively easy problem of what types of very low-risk SRM research should be allowed to continue. There need not be much specific oversight on a project-by-project basis of this low-risk research, such as modelling studies, laboratory investigations or small-scale field experiments.

But over time, as SRM research warrants larger outdoor experiments, the problem of the research posing non-negligible risks would have to be faced. The institutions concerned will have to estimate and evaluate the social risks of doing research, as well as the physical risks. They will need to consider such ethically sensitive questions such as: who is doing the experiment? Who is paying for it? What is the intention? Are the results made public, and how? Do the benefits of increased knowledge outweigh any incurred risks associated with obtaining the knowledge?

Just by developing criteria and protocols for governing SRM research, society will start to learn how we may eventually govern the prospective technology. How will the public consultation occur? How will projects obtain independent advice about whether and how to proceed with research plans? Who has authority to adjudicate? How will the plans and results be made transparent? What are the protections against vested interests having too much sway over decisions? How should governance be coordinated internationally?

Society will have to play a role in governing this research through public engagement and oversight of any activities that aim to modify the environment. Those researchers investigating solar climate engineering will have to find a way to share information and jointly decide on strategy, agree on appropriate goals of the research, decide on research options, monitor the results and make decisions about what to do next. This would all be very good practice for dealing with the problems of climate change in general, free of the immediate socioeconomic impacts that make climate choices so hard.

If the world ever takes a decision to deploy SRM technology, the intervention would overlay one human-induced climatic condition on another. Deployment would not return Earth to a prior, pre-industrial state. Solar climate engineering would essentially 'renovate' the Earth, like renovating an old house that has rotten windows, holes in the roof and a broken furnace. You might never be able to restore it to a former state, but you can probably renovate it to a habitable state. Engaging in SRM research means there has to be dialogue and public engagement about societal values and preferences for the state of the climate. Contemplating solar climate engineering forces us to consider climate change objectives, choices for meeting such objectives, ways to choose among choices and ways to manage those choices that have been made.

The institutions and processes needed to manage this problem do not exist at present (Walker et al., 2009). Perhaps the effort to conduct and manage SRM research will help to support or even build the institutions required to deal with societal and environmental problems more generally. Society will struggle with the management of solar climate engineering research and that struggle will prepare us for governance of possible future deployment choices.

Recognising the likelihood of regional climate interventions

The world already experiences significant disturbance caused by climate change and countries have already deployed local geoengineering interventions designed to alleviate local impacts. Such local interventions may eventually have global implications. SRM research would help countries make wiser decisions about such interventions, prevent unintended global impacts and provide a rational basis for developing governance arrangements.

It seems very likely that extreme weather events made more extreme and more frequent by anthropogenic climate change will catalyse regional climate interventions. Citizens will want and expect governments to do something. Imagine Moscow in a heat wave of 46°C for three months with no end in sight, or the Murray Valley in Australia burning for the entire summer, or tens of thousands dead from heat waves in Delhi or Chicago. Even now Australia's plans for saving the Great Barrier Reef include the possibility of deploying local marine cloud brightening, a form of geoengineering (NAS, 2018). Indonesia currently has a programme of cloud seeding over the ocean to encourage clouds to rain out before they reach Jakarta and cause heavy flooding (Rochmyaningsih, 2013). And the dramatic loss of Arctic sea ice has motivated researchers to think about regional interventions to stop melting (Tilmes et al., 2014). Many of these ideas would also change global weather patterns (Bernstein et al., 2012; MacCracken, 2009).

As impacts from extreme weather events grow, countries will contemplate larger and longer regional geoengineering interventions in response. Eventually either the size or the number of such interventions could have global impacts. Larger and longer regional interventions could eventually become unintentional *global* geoengineering. A highly motivating crisis mentality will result from extreme weather. Unlike global interventions, the motivation to try regional engineering interventions may be very strong and the barriers to deployment may be relatively small. These interventions may start small and resemble local weather modification. But as climate change impacts increase in scale, people will quite likely contemplate larger and multiple interventions that could affect the climate of Earth as a whole. Although we may never know everything about these potential local or regional interventions, we can know more than we do now. The knowledge that comes from SRM research could serve to stop bad ideas or else cautiously prepare us to address unbearable future conditions.

These potential regional responses to extreme conditions also provide an opportunity to grow societal capacity to manage solar climate engineering interventions. Suppose SRM research leads to the design of effective regional climate interventions. Commonly experienced conditions could engender a sense of common purpose among regions with a common problem. This could lead to regional agreements between nations to help each other through climate change related emergencies. Such cooperation might include an agreement to share strategic geoengineering technology and deployment for regional climate interventions. The resulting interactions between nations might include reaching agreement on the goals, scope, methods, assessment approach, treatment of liability issues and financing for extreme weather interventions. Cooperative intervention decisions to ameliorate climate-enhanced natural disasters—especially if supported by research—might help the world improve its capacity to make informed global SRM intervention decisions, as well as to learn-by-doing *before* they face the difficulties of a global intervention (Long, 2013).

Focusing the societal comprehension of climate change

Transparent research on SRM can make it clear that we face existential problems with prospective climate change; it can help people focus on becoming responsible citizens of ***the Anthropocene***.

In this new era, sometimes referred to as the Anthropocene, human activity has become an increasingly dominant influence on climate and the environment. Human dominance of our only habitable planet inherently challenges us to take responsibility for its climate. Many commentators have pointed out that public discussion of solar climate engineering may create a moral hazard in that it could also distract us from the important work of mitigation (Lin, 2013). In fact evidence points to the opposite. Research into SRM might help people understand just how bad the problem is. Consequently, public citizens be more motivated to urge for stronger conventional climate mitigation in order to obviate the need for SRM. The characteristic 'horror' of solar climate engineering communicates to public audiences just how worried some climate scientists are about our future and thus motivates a renewed commitment to mitigation. Some public surveys in the USA show that this may in fact be the case (Braman et al., 2012).

The causes of human-induced climate change have not been introduced intentionally: they have been the inadvertent by-product of the widespread use of fossil fuels which has greatly improved our quality of life. Even though unintentional climate forcing through greenhouse gas emissions is likely to be far more extensive and damaging than that likely to result from an intentional SRM intervention, intentionality does make solar climate engineering problematic. Intention after all is the difference between murder and manslaughter. Yet we cannot avoid intention in the Anthropocene. Since we now know collectively that humans are causing climate change, it really is no longer credible to say that what we continue to do is 'unintentional'. We now have a moral imperative to act responsibly for the state of the environment that we shall leave to future generations. Research into solar climate engineering will bring the concept of intentional management to the fore and perhaps provoke us to take greater responsibility for our actions.

Conclusion

This 'horrible' idea of intentional climate intervention through solar climate engineering goes hand-in-hand with the requirement for taking responsibility for our climate and accepting that our species now has stewardship of the planetary environment. Solar climate engineering creates a tension between the hubris of thinking that humans could manage the problem and the responsibility to try. If climate change unfolds as far and as fast as now seems likely, the risks of not attempting to intervene in this way may exceed the risks of trying to do so. We should not make a decision to *deploy* climate engineering now. But the climate we may have to live with in the future may at some stage demand such a choice from us. This essay supports research into solar climate engineering in order to prepare as best we can for such an eventuality. Research will allow people to make a more informed decision about whether to deploy, especially under potentially dire climatic conditions of the future.

It is possible that grappling with these questions in relation to a research programme will help us to learn the skills for making good decisions in relation to stewardship of the climate and the environment in general. Solar climate engineering is a frightening, and maybe horrifying, idea but it just might become necessary. Between the horrifying and the necessary, research into this technology may create room to grow as a society.

NO: Because it perpetuates the dangerous illusion that a technological fix for climate change is possible (*Rose Cairns*)

Introduction

The argument for research into solar climate engineering has a certain intuitive appeal: given the severity of global climate change we should be exploring all of our possible options and research into solar climate engineering may help us figure out what those options are. I will argue here, however, that it will not. The argument that research on solar climate engineering is necessary and useful rests on at least two important assumptions. First, that research will make it possible to know whether solar geoengineering would 'work' and what the implications of such an endeavour would be (both physically and socially/politically) *before* deploying such a technology at a planetary scale over decade-long time-scales. Second, that it is possible to imagine a future scenario in which such an intervention would be governable in any kind of desirable way. I believe that neither of these assumptions is well founded and that therefore there is no defensible basis for continuing to carry out research in this domain.

Irreducible incertitude

Current knowledge of the possible impacts of solar climate engineering—specifically the idea of injecting aerosols into the stratosphere to reflect sunlight back into space—have come from observation of volcanoes and from climate model simulations. These observations have given rise to the idea that solar climate engineering would be *technically* possible, in the very narrowest sense of the word. On an imaginary planet (i.e. a 'world without politics'—Keith, 2013: 87) a reduction in global *average* temperatures could be achieved. Model simulations tend to agree that some reduction in temperature at the level of global averages could theoretically be achieved. But at the regional and local scales that matter to human life and ecological functioning, for example in the likely local weather disruptions or agricultural impacts of such an intervention, there is much less agreement among the various model predictions. Some models suggest the possibility that solar climate engineering might disrupt the Asian monsoon, create drought over large parts of Africa or have other highly negative outcomes (Vaughan and Lenton,

2011). Of course, unchecked, climate change will also have highly dangerous and uncertain impacts on local weather patterns. Those arguing for more research might frame the future as a 'lose-lose' choice between climate change on the one hand and a geoengineering intervention on the other, arguing that only research can show which would be the 'least worst'.

But the idea that more research will lead to a reduction in controversy, or greater consensus about 'the best' way forward, seems to misrepresent how scientific knowledge operates in society (Sarewitz, 2004). Even a cursory glance at the highly politicised nature of present day global climate knowledge, and the inherent difficulties with attributing a given weather event to either climate change or natural variability (see **Chapter 3**), point to the ways in which claims to scientific knowledge about the precise impacts of any intervention would be endlessly politicised. Claims about the results of solar climate engineering research would always be open to contestation. Indeed, much of the uncertainty and ignorance around the impacts of a solar climate engineering intervention are simply inherent and irreducible. Why should this be?

First, as Hulme has pointed out, beyond those impacts that are currently modelled (such as temperature and rainfall patterns), the full extent of the potential impacts of solar climate engineering 'can barely be imagined, let alone quantified' (Hulme, 2014: 92). The notion that all possible impacts of solar climate engineering could be objectively evaluated through research and presented in terms of quantified risks in order to steer decision making is simply not scientifically rigorous. Under conditions in which neither the outcomes themselves nor the probability of them occurring can be known with any degree of certainty (i.e. when we don't know what we don't know), a more precautionary approach is required. As Stirling cautions, 'it is profoundly irrational and unscientific to seek to represent ignorance as risk' (Stirling, 2008: 100).

Second, even discounting the 'unknown unknowns' and focusing on the 'known unknowns', reducing uncertainty is highly problematic. The primary tool for understanding potential impacts to date has been climate modelling. But whatever claims might be made about their accuracy, because they are mere representations of reality climate models will always be contestable. In order to move toward greater certainty about the technology it therefore needs testing in the real world and at ever-greater scales. However, the problem is that even regional-scale testing of a solar climate engineering intervention would be of limited value in reducing uncertainties about the impacts of a *global* intervention—which is what, by definition, solar climate engineering deployment would need to be.

Indeed, even were a global intervention to take place, controversy and contestation around its impacts would not disappear. As Szerszynski et al. explain, 'the indeterminacy that is endemic to atmospheric and climatic phenomena would mean that the attribution of cause and effect, and of liability and accountability, would be impossible to carry out in any definitive way—even in principle' (Szerszynski et al., 2013: 2811). On the contrary, it would be a matter of intense political debate as to whether any subsequent weather event (e.g. a drought, cyclone, hurricane, etc.) was caused by the intervention or not. This indeterminacy is accepted even by proponents of solar climate engineering research such as David Keith, who writes that 'even if [solar climate engineering] were tested at full scale we would not resolve all of our uncertainties' (Keith, 2013: 63).

The issue is further complicated by the fact that—although it would be possible to detect the impact of an intervention on average global temperatures—it would be highly problematic to distinguish the signal of an intervention from the noise of natural variability in climate at local and regional scales. Solar climate engineering would therefore have to be sustained for *decades* before there could be anything approaching a robust assessment of its impacts (Macmartin et al., 2019; Robock et al., 2010). Thus, as various authors have highlighted, a solar climate engineering intervention would *always* be an experiment at the planetary scale sustained over decades. It is the experimental nature of solar geoengineering that has led to the suggestion that geoengineering research is 'deviant when it comes to the normal process by which science proceeds' (Bunzl, 2010). As Bunzl argues:

> most of science deals with modular phenomena. You can test a vaccine on one person, putting that person at risk, without putting everyone else at risk... even though we have lot of planetary wide goals—like eradicating smallpox—we can test them for untoward effects before full-scale implementation. Not so for geoengineering.
>
> (Bunzl, 2010)

Some proponents of research are sanguine about this blurred boundary between research and deployment and the experimental dimension of the whole enterprise, arguing that solar engineering could be started slowly. In the event that negative impacts were discovered it could then simply be 'turned off'. This is extremely simplistic. Having already embarked upon such a decade-long experiment with the entire planet, it would be difficult if not impossible to change course, even were highly negative impacts discovered. In any case, these too would be contested and difficult to decipher from natural variability over short time-scales. In part there would be strong incentives to continue with solar climate engineering due to the dangerous impacts of the so-called 'termination effect' (Parker and Irvine, 2018).

The idea that such a system could be simply 'switched off' also goes against all current understanding of large scale socio-technical systems, which are notoriously difficult to change due to various processes of 'lock in' (cf. Cairns, 2014a). This is neatly illustrated by the obduracy of the complex, entrenched energy systems which have resulted in our current climate predicament. One only has to consider the kinds of vested interests that would accompany the building of such an enormous physical infrastructure, as well as the social systems (governance institutions, operational personnel, training and so on) that would be co-produced in the process, to understand how complex simply 'switching off the machines' would be. For example, Szerszynski et al. (2013) list just some of the substantial economic opportunities that are likely to be created by any plan to deploy solar climate engineering. These include

> the patenting of specific [solar climate engineering] techniques or classes of techniques; the design of particles for release into the stratosphere; the design of delivery systems; the sourcing and transport of raw materials; the design and implementation of monitoring systems; and the establishment and running of financial schemes of funding and possible compensation.
>
> (Szerszynski et al., 2013: 2814)

Governability

Nearly all those taking part in debates around geoengineering accept that questions of governance are crucially important. However, there is a tendency to act as though questions about the governance of SRM deployment—or more specifically the central question of the *governability* or otherwise of deployment in an imagined future moment—should have no bearing at all on whether or not to invest time and resources into research in the present. In other words, even if all of our current knowledge about social, political and economic realities suggests that solar climate engineering would be ungovernable in any desirable sense, we should nevertheless continue to invest in researching the technology. There are a number of reasons why I do not believe this to be the case and why the argument that we should do research now in order to 'arm the future' (Gardiner et al., 2010) with geoengineering should 'they' (i.e. future generations) need it is neither ethically nor practically defensible.

First, the blurred lines between research and deployment (illustrated by the tendency toward ever-greater scales of research and the ultimately experimental nature of any potential global deployment) make it impossible to discuss research as though it were entirely separate from deployment. Second, developing technologies do not simply emerge 'fully formed' as neutral agents into society, but are rather co-produced with particular social orders (Jasanoff, 2004). So, an important question to ask of those who propose investing in research in this area is what kind of social order would be co-produced with attempts to carry out solar climate engineering of our climate? Would this new world be desirable? For example, inherent to governance of solar climate engineering would be notions of hierarchical steering and expert management by a technocratic elite (Hamilton, 2014). Szerszynski and colleagues discuss what they call the 'social constitution' of solar climate engineering. They argue that embarking on such a programme would be inherently political, in the sense that it would be 'unfavourable to certain patterns of social relations and favourable to others' (Szerszynski et al., 2013: 2811).

Lövbrand and colleagues use the term 'Earth system governmentality' (Lövbrand et al., 2009) to talk about the kinds of social and political relations that would be engendered by geoengineering. They argue that geoengineering, as an effort to centrally monitor and manage the entire Earth system, reinforces the expectation of a singular 'eye of power'. This is what might be called a global governance 'cockpit' from which the Earth system is to be directly controlled and manipulated. Others have pointed to the inevitable secrecy that would accompany such an intervention and the potential for its militarisation (Cairns, 2014b). Szerszynski and colleagues have even argued that it is highly doubtful as to whether such a system would be compatible with a 'pluralist and democratic politics' (Szerszynski et al., 2013). They suggest, to the contrary, that a solar climate engineering programme would be 'compatible with a centralised, autocratic, command-and-control world-governing structure' (ibid p. 2812). Precisely what form this might take is open to conjecture but, as Hamilton puts it,

> [g]iven that humans are proposing to self-engineer the climate because of a cascade of institutional failings and self-interested behaviours, any suggestion that deployment of a solar shield would be done in a way that fulfilled the principles of justice and compassion would lack credibility, to say the least.
>
> (Hamilton, 2014: 21)

Although one might expect the erosion of democracy to be seen as a negative outcome, some prominent voices have actually argued in recent years for the need for more authoritarian and centralised forms of governance given the severity of the environmental threats facing the planet. However, even leaving aside the considerable normative concerns associated with the formation of a **command-and-control** form of global governance (see **Chapter 14**), it is probably naïve to assume that *any* system of global governance could be sustained for the required time span (i.e. centuries) without interruption by wars or ideological disputes. As Schneider surmised more than a decade ago,

> Just imagine if we needed to do all this in 1900 and then the rest of twentieth century history unfolded as it actually did! Would climate control have been rationally maintained, or would gaps and rapid transient reactions have been the experience?
>
> (Schneider, 2008: 3857)

To be fair, most proponents of solar climate engineering research do not *explicitly* suggest that democracy should be 'put on hold', but rather adhere to the ideal of democratic principles of governance characterised by transparency, public participation, and so on (see for example the widely cited Oxford Principles; Rayner et al., 2013). They apparently believe in the possibility that desirable governance structures could be built in advance of any solar climate engineering being undertaken. As a result of this focus on 'good governance' many proponents of solar climate engineering research (although not all—see for example the Arctic Methane Emergency Group) have moved away from explicit reference to 'climate emergency' as a rationale for research. This is precisely because of the problematic nature of emergencies as the basis for good governance. As Sillmann and colleagues summarise,

> declaring an emergency invokes a state of exception which carries many inherent risks: the suspension of normal governance, the use of coercive rhetoric, calls for 'desperate measures,' shallow thinking and deliberation, and even militarization. By definition, emergency situations are extraordinary and exceptional.
>
> (Sillmann et al., 2015: 292)

However, an implicit notion of climate emergency still underlies the rationale for solar climate engineering research, in that it is under the circumstances of a looming or imminent climate crisis that the technology might be 'needed' by people in the future.

Exactly what would constitute such an emergency—particularly one that was claimed to be still imminent rather than actually occurring; who would be authorised to declare such; and, crucially, the ability of a programme of solar climate engineering to address such a situation; none of these things are clear. As Sillmann and colleagues explain, although single events such as hurricanes, droughts or heatwaves might be widely accepted as emergencies in particular geographical locations, no single event or even combination of events would automatically be sufficient to warrant declaration of a *global* climate emergency. Rather it would be 'the global interaction of such events with socioeconomic and political factors, including elements of power and perception, which might eventually determine their designation as global climate emergencies' (Sillmann et al., 2015: 291). Thus, the socioeconomic events that would most likely prompt the declaration of a global emergency might include things such as widespread social unrest or the disruption of supply chains leading to food shortages as a result of extreme weather events. However, given the inherent uncertainties about the regional impacts that would accompany a solar climate engineering intervention it is hard to see how this could be an effective response under such circumstances.

On the contrary, it seems more likely that the regional changes in climate patterns—and all of the uncertainties and contestations around attribution that would accompany these (see **Chapter 3**)—would likely only add further complexity and unpredictability to what would, by definition, already be a chaotic and unpredictable picture. Indeed, rather than recognising solar climate engineering as an effective response to emergency, some authors have explored the potential that the intervention itself could precipitate or exacerbate global crises of various types. For example, Baum and colleagues use scenarios to highlight the extent of the vulnerabilities that would accompany the undertaking of solar climate engineering. They describe a genuine 'worst case scenario' future in which solar climate engineering is started, but then halted due to a global catastrophe (e.g. nuclear war) resulting in global societal collapse (Baum et al., 2013).

Conclusion

One benefit of research about which I *can* agree with proponents of SRM is that through being researched solar climate engineering would 'lose its illusory simplicity' (Keith, 2013: 93). But to those who would embark upon research in this area, I would ask: when would one know enough to know that solar climate engineering wasn't going to be viable? Under what circumstances would a problem be considered sufficiently intractable or serious that researchers collectively invested in this work would decide to abandon it? The answer to these questions would appear to be: never. Rather than encountering the mismatch between the illusory, seductive idea of geoengineering, and its impossibility in the real world as a reason to stop and focus attention elsewhere, the geoengineering mindset would appear to double down or, in Keith's words, to 'find new work-arounds' (ibid p. 93). The campaign against geoengineering (Hands Off Mother Earth: HOME) has also been critical of this kind of dynamic. In a manifesto endorsed by over 100 civil society groups they highlight that since small-scale research cannot reduce uncertainties about global impacts it will 'only serve the purpose of testing hardware and tools to advance research and investments that will then be used to justify "the need" for larger experiments and eventually deployment' (HOME, 2018: 9). While arguments about a 'slippery slope' from research to deployment might be a little overblown, this tendency would appear to be very real.

In summary, I have argued that developing a robust understanding of the physical and sociopolitical impacts of solar climate engineering is effectively impossible without full-scale deployment over decade-long time-scales. This renders research either a futile distraction or else a potentially catastrophic experiment at the planetary scale. Furthermore, any decisions about research cannot ignore discussions about the governance of full-scale deployment. If the deployment of solar climate engineering would seem ungovernable in any desirable sense—as would appear to be the case based on contemporary social and political realities—then the value of continued research is highly questionable. Investing in solar climate engineering research perpetuates the idea that a technological fix for climate change is just around the corner. This is a dangerous illusion.

Further reading

Baskin, J. (2019) *Imagining Geoengineering: Solar Geoengineering, the Anthropocene and the End of Nature*. Basingstoke, UK: Palgrave Macmillan.

Baskin explores the assumptions and imaginaries behind the idea of 'engineering the climate' and discusses why this solution to climate change is controversial and is resisted by so many. He describes three competing imaginaries associated with the technology: an Imperial imaginary, an oppositional Un-Natural imaginary and a conspiratorial Chemtrail imaginary.

Hulme, M. (2014). *Can Science Fix Climate Change? A Case Against Climate Engineering*. Cambridge: Polity Press.

This short book develops a case against solar climate engineering, against both researching it and deploying it. Hulme argues that such a technology is unreliable, undesirable and ungovernable and, as with human germline modification, he calls for a research moratorium.

Keith, D. (2013) *A Case for Climate Engineering*. Cambridge, MA: MIT Press.

Contra to Hulme (2014), Keith argues that after decades during which very little progress has been made in reducing carbon emissions, solar climate engineering technology must be considered responsibly. That doesn't mean we will deploy it, he says, but it does mean that we must understand fully what research needs to be done and how the technology might be designed and used.

Morton, O. (2015) *The Planet Remade: How Geoengineering Could Change the World. The Challenge of Imagining Deliberate Climate Change*. London: Granta.

Morton explores the history, politics and cutting-edge science of solar climate engineering. He weighs both the promise and perils of these controversial technologies and puts them into broad historical context. This clarifies not just the scale of what needs to be done about global warming, but also humanity's relationship with nature.

Reynolds, J.L., Contreras, J.L. and Sarnoff, J.D. (2018) Intellectual property policies for solar geoengineering. *WIREs Climate Change*. 9(2). DOI: 10.1002/wcc.512.

This article reviews the issues and policies relating to the intellectual property of solar climate engineering technologies, including patents and trade secrets, and the possibilities and problems of regulating commercial interests. It argues that innovative policy approaches to intellectual property and data access that are specific to solar geoengineering are needed.

Follow-up questions for use in student classes

1. 'Solar climate engineering should not be ruled out before a serious programme of research has established the benefits and risks of deployment'. To what extent do you agree with this claim?
2. If research into solar climate engineering *is* to continue, what safeguards are necessary to ensure no harm is done through experimentation?
3. It has been suggested that the implementation of a global programme of solar climate engineering would be incompatible with democratic forms of governance. Why might this be so and does it matter?
4. Many organisations, local authorities and even nations (e.g. the UK) have already declared climate change 'an emergency'. Defend or critique the argument that solar climate engineering would be an appropriate and effective response to a climate emergency.

References

Baum, S., Maher, T. Jr. and Haqq-Misra, J. (2013) Double catastrophe: intermittent stratospheric geoengineering induced by societal collapse. *Environment, Systems and Decisions*. 33(1): 168–180.

Bernstein, D.N., Neelin, J.D., Li, Q.B. and Chen, D. (2012) Could aerosol emissions be used for regional heat wave mitigation? *Atmospheric Chemistry and Physics Discussions*. 12: 23793–23828.

Braman, D., Kahan, D.M., Jenkins-Smith, H.C., Tarantola, T. and Silva, C.L. (2012) *Geoengineering and the Science Communication Environment: A Cross-Cultural Experiment*. Yale University: GW Law Faculty Publications and Other Works. Paper No.199

Bunzl, M. (2010) Geoengineering research reservations. Presentation to the annual meeting of the American Association for the Advancement of Science, San Diego, February 2010. Available from: https://sciencepolicy.colorado.edu/students/envs_5000/bunzl_2011.pdf

Cairns, R. (2014a) *Will Solar Radiation Management Enhance Global Security in a Changing Climate?* UCL/University of Sussex/University of Oxford: Geoengineering Governance Research Working Paper Series. Discussion Paper No.16.

Cairns, R. (2014b) Climate geoengineering: issues of path-dependence and socio-technical lock-in. *WIREs Climate Change.* **5**(5): 649–661.

Gardiner, S.M., Stephen, M., Caney, S., Jamieson, D., Shue, H., Franklin, B., Cicerone, R., Fleming, J.R. et al. (2010) Is 'arming the future' with geoengineering really the lesser evil? Some doubts about the ethics of intentionally manipulating the climate system. In S. Gardiner, S. Caney, D. Jamieson and H. Shue, eds. *Climate Ethics: Essential Reading.* New York: Oxford University Press. pp. 1–32.

Hamilton, C. (2014) Geoengineering and the politics of science. *Bulletin of the Atomic Scientists.* **70**(3): 17–26.

HOME (2018) *Hands Off Mother Earth! Manifesto against Geoengineering,* October 2018. Available from: www.geoengineeringmonitor.org/wp-content/uploads/2018/10/home-new-EN-feb6.pdf

Hulme, M. (2014). *Can Science Fix Climate Change? A Case Against Climate Engineering.* Cambridge: Polity Press.

IPCC (2018) Summary for Policymakers. In: *Global Warming of 1.5°C. An IPCC Special Report on the impacts of global warming of 1.5°C above pre-industrial levels and related global greenhouse gas emission pathways, in the context of strengthening the global response to the threat of climate change, sustainable development, and efforts to eradicate poverty* [Masson-Delmotte, V., P. Zhai, H.-O. Pörtner, D. Roberts, J. Skea, P.R. Shukla, A. Pirani, W. Moufouma-Okia, C. Péan, R. Pidcock, S. Connors, J.B.R. Matthews, Y. Chen, X. Zhou, M.I. Gomis, E. Lonnoy, T. Maycock, M. Tignor and T. Waterfield (eds.)]. World Meteorological Organization, Geneva, Switzerland, 32 pp.

Jasanoff, S. (2004) The idiom of co-production. In S. Jasanoff, ed. *States of Knowledge: The Co-Production of Science and Social Order.* Abingdon: Routledge. pp. 1–13.

Keith, D. (2013) *A Case for Climate Engineering.* Cambridge, MA: MIT Press.

Lin, A.C. (2013) Does geoengineering present a moral hazard? *Ecology Law Quarterly.* **40**: 673–712.

Long, J.C.S. (2013) A prognosis, and perhaps a plan, for geoengineering governance. *Carbon and Climate Law Review.* **7**(3): 177–186.

Long, J.C.S. and Shepherd, J. (2014) The strategic value of geoengineering research. In B. Freedman, ed. *Global Environmental Change.* Dordrecht: Springer. pp. 757–770.

Lövbrand, E., Stripple, J. and Wiman, B. (2009) Earth System governmentality: reflections on science in the Anthropocene. *Global Environmental Change.* **19**(1): 7–13.

MacCracken, M.C. (2009) On the possible use of geoengineering to moderate specific climate change impacts. *Environmental Resource Letters.* **4**: 045107.

Macmartin, D.G., Wang, W., Kravitz, B., Tilmes, S., Richter, J.H. and Mills, M.J. (2019) Timescale for detecting the climate response to stratospheric aerosol geoengineering. *Journal of Geophysical Research (Atmospheres).* **124**(3): 1233–1247.

NAS (2018) *A Research Review of Interventions to Increase the Persistence and Resilience of Coral Reefs.* Washington, DC: The National Academies Press.

National Research Council (NRC) (2015) *Climate Intervention: Carbon Dioxide Removal and Reliable Sequestration.* Washington, DC: The National Academies Press.

Parker, A. and Irvine, P.J. (2018) The risk of termination shock from solar geoengineering. *Earth's Future.* **6**(3): 456–467.

Rayner, S., Heyward, C., Kruger, T., Pidgeon, N., Redgwell, C. and Savulescu, J. (2013) The Oxford principles. *Climatic Change.* **121**(3): 499–512.

Ricke, K.L., Morgan, M.G. and Allen, M.R. (2010) Regional climate response to solar-radiation management. *Nature Geoscience.* **3**(8): 537–541.

Robock, A., Bunzl, M., Kravitz, B. and Stenchikov, G.L. (2010) A test for geoengineering? *Science.* **327**(5965): 530–531.

Rochmyaningsih, D. (2013) Is cloud seeding preventing further flooding in Indonesia? [online] *The Guardian*, 19 February. Accessed 18 June 2019. Available from: www.theguardian.com/environ ment/2013/feb/19/cloud-seeding-flooding-indonesia

Sarewitz, D. (2004) How science makes environmental controversies worse. *Environmental Science and Policy.* **7**(5): 385–403.

Schneider, S.H. (2008) Geoengineering: could we or should we make it work. *Philosophical Transactions. Series A, Mathematical, Physical, and Engineering Sciences.* **366**(1882): 3843–3862.

Sillmann, J., Lenton, T.M., Levermann, A., Ott, K., Hulme, M., Benduhn, F. and Horton, J.B. (2015) Climate emergency: no argument for climate engineering. *Nature Climate Change.* **5**(4): 290–292.

Stirling, A. (2008) Science, precaution, and the politics of technological risk: converging implications in evolutionary and social scientific perspectives. *Annals of the New York Academy of Sciences.* **1128** (April): 95–110.

Szerszynski, B., Kearnes, M., Macnaghten, P., Owen, R. and Stilgoe, J. (2013) Why solar radiation management geoengineering and democracy won't mix. *Environment and Planning A.* **45**(12): 2809–2816.

Tilmes, S., Jahn, A., Kay, J.E., Holland, M. and Lamarque, J-F. (2014) Can regional climate engineering save the summer Arctic sea ice? *Geophysical Research Letters.* **41**: 880–885.

Vaughan, N.E. and Lenton, T.M. (2011) A review of climate geoengineering proposals. *Climatic Change.* **109**(3–4): 745–790.

Walker, B., Barrett, S. and 18 co-authors (2009) Looming global-scale failures and missing institutions. *Science.* **325**: 1345–1346.

Part III

On what grounds should we base our actions?

9   Is emphasising consensus in climate science helpful for policymaking?

John Cook and Warren Pearce

Summary of the debate

This debate concerns the public and policy value of communicating the clear scientific consensus about the human role in causing climate change. **John Cook** argues for the importance of frequent and repeated public communication of the overwhelming scientific consensus that human activities cause climate change. Public understanding of the extent of this consensus would seem to act as an important 'gateway belief', a necessary precursor belief for people to recognise the need for climate policies. **Warren Pearce** challenges the significance for policymaking of the Gateway Belief Model, arguing that a preoccupation with consensus messaging plays into the hands of climate change critics by keeping the focus of public discussion on climate science rather than on human values and policy preferences.

YES: Because closing the consensus gap removes a roadblock to policy progress (*John Cook*)

Introduction

The first message that the American public heard about the scientific consensus on climate change was that there was no consensus. In the early 1990s, Western Fuels Association spent over half a million dollars on a marketing campaign with the purpose of 'reposition[ing] global warming as theory (not fact)'. The aim of this campaign was to convince the American public that scientists had yet to form a consensus on whether global warming was happening and human-caused (Oreskes & Conway, 2010).

In 1995, 79 scientists published the Leipzig Declaration, claiming that '[i]t has become increasingly clear that ... there does not exist today a general scientific consensus about the importance of greenhouse warming from rising levels of carbon dioxide' (Powell, 2011: 63). Two years later, in 1997, the Competitive Enterprise Institute created the 'Cooler Heads Coalition', a group of spokespeople disseminating talking points such as 'Many scientists are sceptical of climate change theory' (Powell, 2011: 104). And in 1998, the Oregon Institute of Science and Medicine distributed a petition from tens of thousands of dissenting scientists (titled 'the Global Warming Petition Project') with the purpose of casting doubt on the scientific consensus (Readfearn, 2013).

Over a decade after such industry groups and conservative thinktanks had begun misinformation campaigns against the consensus, scientists began to quantify and communicate the degree of scientific agreement that human activities cause global warming (hereafter referred to as 'the consensus'). The first study found no dissenting papers in 'global climate change' research published from 1993 to 2003 (Oreskes, 2004). Subsequent studies have found 97% agreement about anthropogenic global warming (AGW) among published climate scientists (Doran & Zimmerman, 2009) and within relevant climate papers (Cook et al., 2013). A synthesis of consensus studies found that among published climate scientists, the level of agreement on human-caused global warming ranged from 90 to 100%, with multiple studies converging around 97% (Cook et al., 2016).

Despite the robust body of research establishing overwhelming consensus, misinformation that emphasises scientific *dis*agreement about climate change continues to be prolific and persistent (Figure 9.1). An analysis of syndicated conservative columns about climate change from 2007 to 2010 found that the most common argument was 'there is no scientific consensus' (Elsasser & Dunlap, 2013). The most shared climate change-related article during the 2016 USA election was a conspiratorial article promoting the Global Warming Petition Project (Readfearn, 2016). In 2017, the Heartland Institute mailed copies of the book *Why Scientists Disagree About Global Warming* to 25,000 science teachers across the USA (Worth, 2017). In late 2018, President Trump argued that on climate change 'you have scientists on both sides of the issue' (Associated Press, 2018). Why has so much money and effort gone into undermining the scientific consensus on climate change?

The empirical consensus on consensus messaging

Opponents of climate action have shown such an intense focus on discrediting the consensus because this form of misinformation is effective in delaying climate policy. Early industry and political market research found that anti-consensus messaging decreased public support for climate policy. A 1998 report by the American Petroleum Institute concluded that informing people that some scientists did not believe humans were causing global warming made them more likely to oppose the 1997 Kyoto Protocol. Similarly, a strategy memo in 2002 by political strategist Frank Luntz offered the following advice to Republican politicians engaged in public debate about climate policy:

> Voters believe that there is no consensus about global warming in the scientific community. Should the public come to believe that the scientific issues are settled, their views about global warming will change accordingly. Therefore, you need to continue to make the lack of scientific certainty a primary issue in the debate.
>
> (Luntz, 2003: 137)

Market research from political strategists and by industry bodies into the influence of consensus misinformation have been subsequently confirmed by social science research. Misinformation citing the Global Warming Petition Project has been shown in the United States to lower perceived consensus and climate policy support amongst the public, as well as confound belief in accurate climate information (Cook et al., 2017; van der Linden et al., 2017a). Testing of a range of misinforming statements about climate change found that an attack on scientific consensus was the most effective in reducing public acceptance of climate change (van der Linden et al., 2017a). Communicating even modest amounts of dissent amongst experts has been shown to be effective in reducing support for environmental policy (Aklin & Urpelainen, 2014).

Why is misinformation targeting the scientific consensus on climate change so potent? When it comes to complicated scientific topics such as climate change people tend to

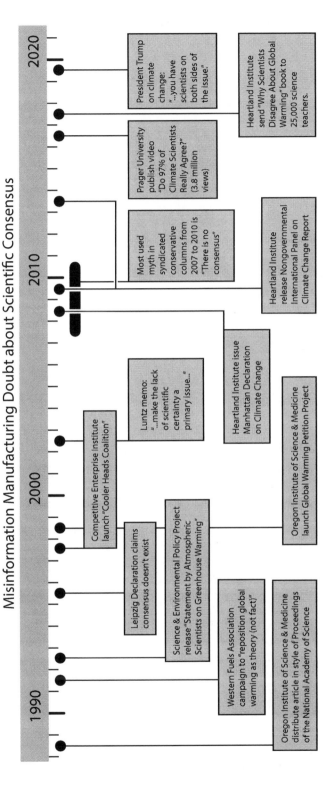

Figure 9.1 Timeline of selected misinformation campaigns targeting the scientific consensus on climate change

rely on heuristics such as expert opinion. When people think experts disagree about human-caused global warming, they are less likely to think global warming is happening or that it is human-caused. Conversely, when people become aware of the scientific consensus they also become more accepting of human-caused global warming. This dynamic has been captured by the Gateway Belief Model (GBM), which proposes that a shift in perceived extent of scientific consensus causes a shift in attitudes and that this subsequently changes support for public action.

The importance of perceived consensus and the efficacy of consensus messaging has been demonstrated in a large and growing number of correlational and experimental studies. Table 9.1 summarises research into consensus messaging or associations between public perception of consensus and other climate attitudes. While there are many different beliefs and attitudes about climate change, survey analysis has shown that the five most important beliefs are that global warming is happening, it's human-caused, the impacts are serious, the experts agree on these first three points and that the problem can be solved. Public perception about expert consensus has been found in multiple studies to be a 'gateway belief' influencing the other four key climate beliefs. The studies listed in Table 9.1 have been categorised as either correlational, using survey data to demonstrate associations between perceived consensus and other climate attitudes, or else experimental, employing randomised experiments to establish causal links between consensus messaging and beliefs about climate change or support for climate policy.

Across this substantial and growing body of research there is diversity in the focus of each analysis. Some studies examine the influence of perceived consensus on beliefs about climate change, while others also look at the influence on policy support. Nevertheless, the consistent picture emerging is that accurate communication of the scientific consensus is effective in increasing public acceptance of human-caused climate change, which positively influences policy support. The overwhelming majority of studies are consistent with GBM. Further, most studies that measure political ideology find that consensus messaging has a depolarising effect, producing greater belief change among political conservatives. While the studies are heavily skewed towards USA citizens, support for GBM has been replicated in a number of countries (e.g. Australia, Japan, New Zealand, UK and USA), while Hornsey et al. (2016) features a meta-analysis of data from over 30 countries. Figure 9.2 demonstrates that the overwhelming majority of published research studying perceived consensus or consensus messaging is either consistent with the Gateway Belief Model or else supports the efficacy of consensus messaging.

The Gateway Belief Model has also been replicated in other scientific topics such as vaccination (van der Linden et al., 2015a), genetically modified food (Kerr & Wilson, 2018b), young-Earth creationism (Tom, 2018) and less politically loaded topics such as water pollution (Dunwoody & Kohl, 2017). The strong evidence supporting consensus messaging on the issue of climate change, plus its generalisability to different countries and other scientific topics, powerfully challenges the assertion that GBM is based on scant empirical evidence (Pearce et al., 2017a).

Addressing objections to consensus messaging

Despite the strengthening empirical basis establishing the efficacy of consensus messaging, and the well-documented campaigns against the consensus waged by opponents of climate policy, some commentators argue that scientists should refrain from emphasising

Table 9.1 Studies into perceived consensus/consensus messaging/GBM with respect to climate change ('Support' column indicates whether or not the study supports GBM or consensus messaging)

	Author (year)	Study type	Country	Finding	Support
1	Malka et al. (2009)	Correlational	USA	Perceived consensus mediates association of knowledge with climate concern among Democrats and Independents who trust scientists.	Y
2	Ding et al. (2011)	Correlational	USA	Low perceived consensus is associated with lower climate beliefs and lower policy support.	Y
3	Lewandowsky et al. (2013)	Experimental	Australia	Consensus messaging increases acceptance of AGW.	Y
4	Rolfe-Red Rolfe-Redding et al. (2011)	Correlational	USA	Perceived consensus predicts climate beliefs and attitudes among Republicans.	Y
5	McCright et al. (2013)	Correlational	USA	Perceived consensus affects policy support, mediated by global warming beliefs.	Y
6	Aklin and Urpelainen (2014)	Experimental	USA	Modest amounts of scientific dissent undermine public support for environmental policy.	Y
7	Bolsen et al. (2014)	Experimental	USA	Consensus messaging reduces partisan differences on behavioural intent and belief in AGW.	Y
8	van der Linden et al. (2014)	Experimental	USA	Consensus messaging (in pie-chart form) reduces partisan difference in perceived consensus.	Y
9	Myers et al. (2015)	Experimental	USA	Consensus messaging is equally effective among liberals and conservatives.	Y
10	van der Linden et al. (2015b)	Experimental	USA	Increasing perceived consensus is significantly and causally associated with climate beliefs, which predicts increased policy support.	Y
11	Cook and Lewandowsky (2016)	Experimental	Australia USA	Consensus messaging reduces partisan differences on belief in AGW for Australians. It increases partisan differences for Americans but still have an overall positive effect on belief in AGW.	Y
12	Deryugina and Shurchkov (2016)	Experimental	USA	Consensus messaging increases acceptance of climate change and human causation.	Y
13	Hamilton (2016)	Correlational	USA	Acceptance of AGW correlates with perceived consensus.	Y
14	Hornsey et al. (2016)	Correlational	USA, UK, Australia, 30 European countries	Perceived consensus is a strong predictor of belief in climate change (stronger than cultural cognition).	Y
15	Schuldt and Pearson (2016)	Correlational	USA	Perceived consensus is associated with mitigation support for both whites and non-whites.	Y

(Continued)

Table 9.1 (Cont.)

	Author (year)	Study type	Country	Finding	Support
16	Brewer and McKnight (2017)	Experimental	USA	Comedy segment about consensus has strongest effect on belief in climate change among participants with low interest in the environment.	Y
17	Cook et al. (2017)	Experimental	USA	Consensus messaging neutralises polarising influence of misinformation.	Y
18	Dixon et al. (2017)	Experimental	USA	Consensus messaging does not produce significant effects (including no backfire effect among conservatives).	Neutral
19	van der Linden et al. (2017a)	Experimental	USA	Consensus messaging reduces partisan differences on perceived consensus.	Y
20	Bolsen and Druckman (2018a)	Experimental	USA	Consensus messaging backfires with conspiracy theorists, but consensus messaging coupled with belief validation increases acceptance of AGW among conspiracy theorists.	Neutral
21	Bolsen and Druckman (2018b)	Experimental	USA	Consensus message increases perceived consensus with indirect effect on belief in AGW and policy support.	Y
22	Harris et al. (2019)	Experimental	UK	Consensus messaging increases perceived consensus and climate beliefs.	Y
23	Kerr and Wilson (2018a)	Correlational	New Zealand	Perceived consensus does not predict later personal climate beliefs.	N
24	Kerr and Wilson (2018b)	Experimental	New Zealand	Consensus messaging increases perceived consensus with indirect effect on belief in AGW.	Y
25	Kobayashi (2018)	Correlational, Experimental	Japan	Perceived consensus predicts climate beliefs. Consensus messaging increases climate beliefs through perceived consensus.	Y
26	Tom (2018)	Correlational	USA	Misconception about consensus is one of the most important factors in predicting scientifically deviant beliefs.	Y
27	van der Linden et al. (2018b)	Correlational	USA	Perceived consensus did predict later personal climate beliefs.	Y
28	Zhang et al. (2018)	Experimental	USA	Consensus messaging is most effective in conservative parts of the USA.	Y
29	Goldberg (2019)	Experimental	USA	Consensus messaging reduces partisan differences on perceived consensus.	Y
30	Ma et al. (2019)	Experimental		Consensus messaging produces reactance among conservative dismissives.	N
31	van der Linden et al. (2019)	Experimental	USA	Consensus messaging increased climate beliefs and attitudes, which were associated with increases in support for action. Conservatives showed greater belief updates.	Y

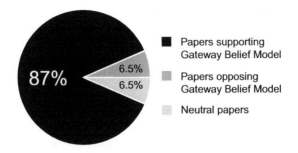

Figure 9.2 Consensus on consensus messaging: the proportions of studies applied to climate change (n = 31) supporting or opposing consensus messaging/Gateway Belief Model

the scientific consensus on climate change. While a variety of arguments have been deployed against consensus messaging, these arguments show little coherence. For example, some argue that consensus messaging does not work (Kahan, 2015), while also arguing that consensus messaging has been successful and is no longer necessary (Kahan, 2016). Both arguments are refuted by a battery of nationally representative surveys of the USA public, finding that the public perception of the scientific consensus has been steadily increasing over the last decade (see Figure 9.3; Cook et al., 2018). Nevertheless, there is still a significant 'consensus gap' between the American public's

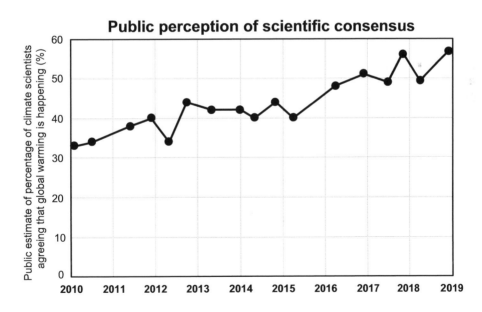

Figure 9.3 Public estimate of the percentage of climate scientists agreeing that humans are causing global warming, derived from USA national representative surveys

Source: Cook et al., 2018.

perception of how strong the consensus is and the 97% consensus evident from the studies cited earlier. Recent surveys have found that only 13% of Americans were aware that the actual scientific consensus is over 90%.

Another argument against consensus messaging is that it detracts from policy discussion (Pearce et al., 2015, 2017a). However, constructing a choice between consensus messaging and political deliberations is a false dichotomy. Consensus messaging *complements* rather than competes with policy discussion (Cook, 2017; Cook et al., 2018). Further, Pearce et al. (2017a) assert that consensus messaging is not a magic bullet to policy progress. However, this argues against a straw man—no one asserts that an accurate perception of consensus or indeed knowledge of any single scientific fact is a turnkey for policy progress. On the contrary, communication experts are in full agreement that consensus messaging is not a magic bullet (Maibach & van der Linden, 2016). The purpose of consensus messaging is not to resolve policy questions. Communicating scientific consensus on climate change simply establishes humanity's role in causing warming—with the implication that we need to mitigate our actions, as demonstrated in the empirical results listed in Table 9.1. Closing the consensus gap removes a roadblock to policy progress.

Pearce et al. (2015) argue that consensus messaging places scientists in the line of fire from opponents of climate action. As Figure 9.1 demonstrates, the status of the scientific consensus has been under attack from political adversaries since the early 1990s, over a decade before the first study quantifying the consensus. Consensus messaging did not precipitate attacks on consensus. Self-censorship will not deter opponents of climate action from attacking the scientific consensus and attempting to undermine public trust in climate science. Indeed, the idea that self-censoring will protect scientists from denialist attacks is symptomatic of a broader malaise afflicting the scientific community—a reluctance to properly engage with the issue of science denial and misinformation. Refusing to address the pernicious impacts of misinformation is dangerous and harmful. Misinformation fosters misconceptions (Ranney & Clark, 2016), reduces policy support (van der Linden et al., 2017a), cancels out accurate information (McCright et al., 2016) and polarises the public (Cook et al., 2017). Misinformation also has a chilling effect on the scientific community, influencing how they report their scientific results (Lewandowsky et al., 2015).

Conclusion

In order to recognise why it is important to communicate the scientific consensus that humans cause global warming, and why it is dangerous to not do so, one needs to consider the purpose of the misinformation that is targeting the consensus. Opponents of climate action target the consensus because this strategy is successful in inhibiting policy discussion—through fostering misconceptions about expert opinion and thereby decreasing public support for climate action because the public believes that there is still a scientific debate. The consensus gap is a roadblock that is delaying climate policy development and implementation, a so-called 'lever for inaction' (Cook et al., 2018). Consensus messaging removes this roadblock. Rather than detract from deliberations about climate policy, consensus messaging in fact enables such discussion.

Consequently, a clear-eyed, evidence-based response to misinformation is required. Science denial is a complex phenomenon that requires multidisciplinary solutions. Scientific messages (including consensus information) can be safely communicated without fear of being neutralised or politicised by misinformation if recipients are inoculated with explanations of the misleading techniques employed by denialists (Cook et al., 2017; van der Linden et al., 2017a). To remain silent about the extent of the scientific consensus and leave consensus misinformation unopposed will result in deepening public confusion about climate change, thereby eroding public support for climate action.

More broadly, it is unfortunate that simplistic false dichotomies abound in discussions of science communication. One such example, tangentially related to this issue, is the 'debate' between the ***information deficit model*** (which proposes that public scepticism about climate change is due to a lack of information) and *cultural cognition* (which proposes that risk perceptions are informed by peoples' values—Kahan & Carpenter, 2017). Both theories contribute useful insights into how to effectively communicate climate change. A constructive approach integrates elements from both models across a variety of social and cognitive contexts (van der Linden et al., 2017b). Similarly, the array of communication strategies available to scientists and communicators should not be reduced to a single approach. Effectively reaching a diverse audience requires a holistic, nuanced strategy incorporating multiple, complementary frames. Climate communication is not a zero-sum game.

Arguments that discourage scientists and social scientists from communicating the consensus play into the hands of lobbyists and politicians who seek to delay climate policy. Failure to communicate the scientific consensus leaves the public vulnerable to harmful misinformation, not only eroding public support for climate policy, but also decreasing public levels of climate literacy (Maibach & van der Linden, 2016). By ignoring the growing body of empirical research supporting the efficacy of consensus messaging about human-caused global warming and reducing climate communication issues to simplistic false dichotomies, opponents of consensus messaging inadvertently contribute to the very outcome they seek to avoid—reduced public support for climate policies.

NO: Because consensus is narrow and human values are more important for policy-making (*Warren Pearce*)

Introduction

There is a scientific consensus that human activity is responsible for most of the observed rise in global temperature since pre-industrial times—hereafter referred to as 'the climate consensus'. Successive studies have attempted to quantify this consensus, typically finding agreement in over 90% of academic papers or scientists surveyed, with the figure of 97% being the most widely quoted (Pearce et al., 2017a). Some science communication researchers argue that public knowledge of these findings, particularly in the USA, has been suppressed by the actions of climate policy opponents who have criticised and cast doubt upon climate science. By countering such misinformation, science communicators seek to increase public awareness of the climate consensus and thereby increase public support for climate policy. Notable efforts in this regard include The Consensus Project,[1] which places great focus on the 97% figure and has successfully inculcated the idea of climate consensus in popular culture—for example, in TV programmes such as *The John Oliver Show*[2] and in politics with its inclusion in the Obama Administration's climate change communication.[3] Some climate change communication campaigns, again notably in the USA, also seek to 'call out' politicians who cast doubt on the consensus—for example, through Climate Nexus's daily 'Denier Roundup' newsletter section.

In short, the maintenance and defence of the climate consensus is an important concept running through the heart of how climate change has been framed in the last three decades. This framing rests on three assumptions: the content of the climate consensus is important; public awareness of the climate consensus is consequential; acceptance of the climate consensus provides a necessary precursor for political progress.

I argue here that contrary to this established framing of climate change, all three of these assumptions are fundamentally flawed:

1. the content of the climate consensus is remarkably narrow and is insignificant relative to the most crucial issues within climate change;
2. knowledge of the climate consensus is a poor way to measure public attitudes towards climate change;
3. convergence of interests, rather than agreement over technical knowledge, is the most important pre-requisite for political cooperation.

I summarise the evidence against emphasising climate consensus in each of these three domains, before offering some concluding reflections on the relationship between science and politics and some options for deeper democratic engagement with climate change.

The climate consensus is strong, but narrow

There is little or no crossover in climate change between the issues where scientific consensus exists and the issues that matter most for informing societal responses. The widely messaged climate consensus does not denote agreement about matters of importance, only that the subject of the consensus—that humans cause global warming—is narrow enough to gain agreement upon. The truth is that the climate consensus relates to only a small subset of the broad sweep of issues assessed and reported on by the IPCC. The consensus tells us nothing about the future of climate change, such as human and non-human impacts, policy options or the range of human values and cultures which interact with local climates. In matters such as these, there is little sign of scientific (or broader public, sometimes called social) consensus. For example, there is ongoing disagreement regarding the volume of greenhouse gases that can be emitted before the **Paris Agreement** target of 1.5°C or 2°C of warming is breached; the economic value that is attributed to future climate-related losses; and the contribution of Antarctic ice melt to sea-level rise. The important question to ask is not 'which knowledge are we the most certain about', but rather 'which knowledge matters the most'.

This point is important for three reasons. First, by focusing on those areas of knowledge that matter most, we can better discuss and delineate societal responses to climate change. These responses necessarily entwine scientific knowledge with normative views about the kind of world in which we want to live in the future. One long-standing example has its origins in climate economics, where debates have raged over the appropriate discount rates to be applied to future impacts (see **Chapter 5**). While operating primarily at a technical level within economics, this debate has opened up important normative issues over how we imagine future uncertainties about climate change and the value we ascribe to the world of our grandchildren compared to the world of today. The question of discount rates also highlights a key dynamic within climate change debates: the vast time-scales involved do not sit easily with our 'common-sense' horizons of understanding (Jasanoff, 2010). Exploring these issues is crucial to advancing public engagement with climate change and determining political responses. Yet it is crucial to note that with the question of discount rates, as in other key climate debates, *no objective answer exists and no scientific consensus is likely, or even possible.* This makes such value-based questions no less important to discuss. In fact, it makes them more important to be exposed to debate.

Second, keeping the spotlight trained on one narrow part of physical science encourages further public debate regarding the science, rather than about how to respond to climate change. Starting with science provides a barrier to public deliberation on climate change. It grounds debates in technical details rather than beginning from a values-based discussion about the different kinds of futures we can imagine for our societies, both utopian and dystopian. Presenting climate policy as a *fait accompli* flowing from a scientific consensus encourages critics to engage in relatively inconsequential technical exchanges, rather than focus on questions of values and politics. Not only does this restrict broader public engagement, it also places science in the line of political fire. This exposes scientific knowledge to political attacks that it is ill-equipped to resist (Pearce et al., 2017a, 2017b).

Third, the veneration of climate consensus sets unreasonably high expectations regarding what scientific knowledge can offer to societies, implying that this 'gold standard' of agreement is the benchmark by which scientific inputs into policy should be measured in the future. Achieving durable and robust consensus assessments, such as those undertaken by the IPCC, requires huge investments of resources that have considerable opportunity cost. While consensus statements may please those who value the erasure of ambiguity—such as some climate science communicators—such statements also likely encourage the exclusion of other important, but less certain, scientific findings. It presents science as 'speaking with one voice' rather than a more accurate representation of a diverse community of scientists with different ideas and disciplinary approaches.

Knowledge of climate consensus is a poor measure of public attitudes

Since the 1970s, surveys have sought to discover people's knowledge of scientific facts. A perceived decline in public trust in science in the 1980s and 1990s brought about a renewed effort to boost the public understanding of science, seeking to fill perceived gaps in knowledge. However, it soon became clear that starting public engagement from an assumption of public ignorance was a failed strategy (Wilsdon and Willis, 2004). This information deficit approach to science communication did nothing to arrest a series of public science controversies including 'mad cow disease' (BSE, or bovine spongiform encephalopathy), the measles/mumps/rubella vaccine and genetically modified crops. This prompted a shift in public engagement during the 2000s from 'deficit to dialogue' as it became clear that simply increasing or accelerating the flow of scientific information fell way short of addressing the disagreements over values and assumptions that so often underpin scientific controversies (Trench, 2008).

Despite these well-documented trends in public engagement and science communication, climate consensus communicators appear stuck in the 1990s, assuming that the public are ignorant and require education to make them more amenable to climate policy proposals. There is copious literature demonstrating that political values and identities are a key driver of attitudes to climate change, particularly in the USA (for example, Kahan et al., 2011). While there is some limited evidence that informing USA

Table 9.2 American voters' opinion on the importance of climate change, compared to knowledge of scientific consensus

Opinion	%
Agree schools should teach causes, consequences and potential solution to global warming *	78
Think global warming is happening *	73
Think US should participate in Paris Agreement **	69
Think global warming is important personally *	72
Worried about global warming *	69
Understand that over 90% of climate scientists think human-caused global warming is happening *	13

Source: Pearce et al. (2017b).

conservatives about the scientific consensus may have a small effect in increasing support for climate policy, this evidence comes mostly from lab-based studies in controlled conditions. In the real world, citizens constantly come across messages which could complicate or contradict a bald consensus message, especially if they rely on media sources that are aligned with their own political identity.

Even if one did believe that science communicators should prioritise deficit over dialogue, the evidence from survey data suggests that there is already significant public understanding of, and concern about, climate change. There is also evidence of support for a range of policy measures aimed at addressing the issue. Even in the USA, the country where there is perhaps the greatest scepticism about climate science, opinion polling from 2017 suggests a clear majority position that climate change is real, personally important and worrisome (see Table 9.2) (Pearce et al., 2017b). This is despite only 13% of Americans knowing that the scientific consensus on human-caused global warming is endorsed by over 90% of climate scientists. This survey data provides no credible evidence that a deficit of public knowledge regarding scientific consensus holds back public knowledge and concern about climate change.

Social media studies are also starting to contribute to understanding in this area (see **Chapter 15**). One study shows that local temperature anomalies affect the rate of tweets posted about climate change (Kirilenko et al., 2015). This suggests that, whether or not such anomalies can be scientifically linked to human activities, the idea of climate change has become well-embedded in societies and is frequently discussed on the occasion of unusually warm or cold days. In short, there is good evidence that climate change is a widely known and concerning phenomenon amongst the public. Choosing instead to focus on the size of the 'consensus gap' as a measure of public understanding is obtuse at best. It is largely irrelevant to the much bigger questions concerning the political responses to climate change.

Convergence of interests, not technical agreement, is the precursor for policymaking

To understand why establishing a narrowly drawn climate consensus makes a poor basis for policymaking, we can turn to the academic literature on evidence-based policymaking. The relationship between the production of scientific knowledge, and its use as evidence for policymaking, is often assumed to be linear. That is, once scientific knowledge is accepted as 'fact' then politicians and policymakers can move on to the next stage of responding to that fact. This 'linear model' of science into policy has more to do with ideals of how policy-making *should* work, rather than looking at the reality of how policymaking *does* work.

One way to think about this is in terms of politicians' *supply and demand* for scientific knowledge (Cairney, 2016). In the linear model, the focus is on the supply of scientific advice and the desire to develop the strongest, most robust example of that advice possible. For climate change, this has often been expressed through efforts to forge scientific consensus, an ever-present aim of the IPCC since its first report in 1990. Under such a supply-side model, scientific experts such as those shaping the IPCC are granted a primary role in problem definition. While we now take for granted that climate change is a global problem with global solutions, up until the late 1980s environmental problems linked to carbon dioxide emissions were typically framed as *local* in character (Miller, 2004). The shift to considering climate in terms of a *global* system has largely been driven by the scientific community, strongly influenced by the emergence of global climate models. This development also aided the building of the scientific consensus, which rests heavily on the calculation of a single figure for global average temperature change.

This change in problem definition from local to global has been *supplied* by scientists, but was not necessarily *demanded* by politicians. The latter have unsurprisingly found it challenging to reach international, implementable agreements to reduce greenhouse gas emissions when their political influence remains focused on nation-states (see **Chapter 12**). A more realistic view of policymaking is through the demand side. Politicians have multiple issues that they want to consider and address based on the policy preferences and underlying values upon which they were elected to office. These are the issues that politicians will seek evidence for, not those issues about which there happens to be the greatest scientific consensus.

This does not mean that political progress in climate change is impossible. It just means that it does not need to be rooted in the climate consensus. For example, in late 2018 the Sunrise Movement were very successful in raising climate change up the political agenda in the United States, leading many Democrat presidential candidates to back the Green New Deal policy programme. A key narrative underpinning the Movement was that only 12 years remained to address climate change, originating from a media report of the IPCC Special Report on 1.5°C (Watts, 2018). Despite its scientific origins, the '12 years narrative' has attracted a lot of criticism from climate scientists, who have described it as a misleading representation of the IPCC report and an unhelpful framing (Asayama et al., 2019). Yet despite this overt *dissensus* about the technical nature of the claim, it has proved extraordinarily effective at provoking new political conversations about climate change between a broad array of actors in American civil society.

Contrast this with the communication efforts made by the Obama Administration, who placed the climate consensus message at the heart of their communication campaign. They used it as a means to denounce Republican politicians who had dismissed the reality of the climate consensus. Such dismissals of science may be frustrating, but should not be assigned unwarranted importance by climate policy advocates. Consensus over particular facts is *not* a requirement for political cooperation; rather it is the *convergence of interests* that is essential (Rescher, 1995). Such interests can converge from unlikely places. For example, Republican Governor of Texas Rick Perry (later Secretary of State for Energy) has never accepted the climate consensus and yet heavily promoted wind power generation while in office. In the UK, researchers have shown how reframing climate change around nature conservation or national pride can build conservative support for climate policy (Pearce et al., 2017a; Whitmarsh and Corner, 2017).

Conclusion

Writing in 1979, the influential scholar of knowledge and society Carol Weiss noted that much research that gains wide currency is oversimplified or inadequate. This leads not to the enlightenment of societies, but to their 'endarkenment' (Weiss, 1979). Four decades on, the promulgation of the climate consensus is a prime example of this enduring malaise. Not because the scientific consensus about the human causes of climate change is wrong, but because it splits the issue of climate change asunder from its local, human context. New, localised forms of public policy dialogue are emerging, particularly around issues such as air pollution and human health. These are local environmental issues which represent important policy areas where interests can converge with the emissions reduction aims of climate policy. More broadly, the UK's My2050 engagement project[4] invited citizens to explore the trade-offs inherent in climate policy and to submit their own ideas for the future mix of energy generation. The ability to reflect on such trade-offs moves beyond the technical question of different energy technologies' carbon emissions to include fundamental questions about the values that underpin public energy choices and what people do and do not value about the world they live in.

For anyone talking about climate change in the public sphere, the choice is this: is your priority the lionisation of scientific knowledge as the basis for political action; or is it nurturing fertile ground for political cooperation? In an ideal world, one might wish these to be one and the same thing. Sadly, we do not live in such a world. If one thinks the most important challenge is to force all unwilling politicians and resistant publics to accept the framing and content of scientific assessment reports, then that is one choice. I suggest a different path: setting aside science-based purity tests in favour of finding common ground for policy cooperation. For some, relegating climate science to the background of political discussion may appear a retrograde step. However, if we are *really* trying to make policy progress then it is nothing less than essential. Continuing the political focus on the strength of a narrowly drawn climate consensus only plays into the hands of climate change critics. It allows them to indulge in arcane technical issues about the true strength of the consensus rather than engaging our publics in the messy, unsatisfying and ultimately unavoidable business of policy progress through political dialogue and compromise.

Acknowledgements

Warren Pearce acknowledges the support of the Economic and Social Research Council Future Research Leaders programme for the Making Climate Social project (ES/N002016/1).

Further reading

Cook, J., van der Linden, S., Maibach, E. and Lewandowsky, S. (2018). *The Consensus Handbook: Why the Consensus on Climate Change is Important*. Fairfax, VA: George Mason University. Available at: https://climatechangecommunication.org/wp-content/uploads/2018/03/Consensus_Handbook-1.pdf

This handbook provides a brief history of the consensus on climate change. It summarises the research quantifying the level of scientific agreement on human-caused global warming and examines what the public thinks about the consensus, and the misinformation campaigns that have sought to confuse people. It answers some of the common objections to communicating the consensus.

Leiserowitz, A., Maibach, E., Roser-Renouf, C., Rosenthal, S. and Cutler, M. (2017) *Climate Change in the American Mind: May 2017*. Yale University and George Mason University. New Haven, CT: Yale Program on Climate Change Communication.

This report is based on findings from a nationally representative survey of American citizens conducted in 2017 by the Yale Program on Climate Change Communication and the George Mason University Center for Climate Change Communication. It reveals American citizens' attitudes to climate science and climate consensus, to climate risks and to climate policies.

Machin, A. (2013) *Negotiating Climate Change: Radical Democracy and the Illusion of Consensus*. London: Zed Books.

Fierce disagreement still exists over the best way to tackle climate change or, indeed, whether it should be tackled at all. In this book, Machin draws on radical democratic theory to show that such disagreement does not have to hinder collective action; rather, democratic differences are necessary if we are to have any hope of acting against climate change.

Sarewitz, D. (2011) Does climate change knowledge really matter? *WIREs Climate Change*. **2**(4): 475–481.

This article explains how climate science and climate policy have become tightly linked. The scientific consensus is supposed to provide the factual basis for action on climate, while the policy approach to dealing with climate is supposed to be guided by this consensus. Sarewitz explains why new policy paradigms are needed that can enfranchise more diverse political constituencies and which do not lean so heavily on defending 'the climate consensus'.

van der Sluijs, J.P., van Est, R. and Riphagen, M. (2010) Beyond consensus: reflections from a democratic perspective on the interaction between climate politics and science. *Current Opinion in Environmental Sustainability*. **2**(5–6): 409–415.

This article distinguishes between three strategies to deal with scientific uncertainties in interfacing climate science and policy: quantify uncertainty; build scientific consensus; and openness about ignorance. The IPCC has been guided by the consensus approach, but there are weaknesses to this approach. A case study from the Netherlands shows how the consensus approach can hinder the necessary political climate debates.

Follow-up questions for use in student classes

1. Why have opponents of climate action focused on casting doubt on the scientific consensus? Is this a potentially effective strategy in delaying climate policy?
2. Should politicians who hold fringe views about scientific issues be ostracised by other politicians and the media?
3. Should more be done to ensure that only correct scientific information is ever published in the media? Why? How? And by whom?
4. Under what different circumstances might societies wish to 'open up' or 'close down' political questions about responding to climate change?
5. What strategies might be employed to neutralise the influence of misinformation?

Notes

1 See: http://theconsensusproject.com/ Accessed 2 June 2019.
2 See: www.youtube.com/watch?v=cjuGCJJUGsg Accessed 2 June 2019.
3 For a full list of media coverage see: web.archive.org/web/20190122175039/www.skepticalscience.com/republishers.php?a=tcpmedia Accessed 2 June 2019.
4 See: http://2050-calculator-tool.decc.gov.uk/#/home Accessed 2 June 2019.

References

Aklin, M. and Urpelainen, J. (2014) Perceptions of scientific dissent undermine public support for environmental policy. *Environmental Science & Policy.* **38**: 173–177.

Asayama, S., Bellamy, R., Geden, O., Pearce, W. and Hulme, M. (2019) Why setting a climate deadline is dangerous. *Nature Climate Change.* **9**(8): 570–572.

Associated Press (2018). Read the transcript of AP's interview with President Trump [online]. *Associated Press* 27 October [viewed 2 June 2019]. Available at: https://apnews.com/a28cc17d27524050b37f4d91e087955e

Bolsen, T. and Druckman, J.N. (2018a) Validating conspiracy beliefs and effectively communicating scientific consensus. *Weather, Climate, and Society.* **10**(3): 453–458.

Bolsen, T. and Druckman, J.N. (2018b) Do partisanship and politicization undermine the impact of a scientific consensus message about climate change? *Group Processes & Intergroup Relations.* **21**(3): 389–402.

Bolsen, T., Leeper, T.J. and Shapiro, M.A. (2014) Doing what others do: norms, science, and collective action on global warming. *American Politics Research.* **42**(1): 65–89.

Brewer, P.R. and McKnight, J. (2017) 'A statistically representative climate change debate': satirical television news, scientific consensus, and public perceptions of global warming. *Atlantic Journal of Communication.* **25**(3): 166–180.

Cairney, P. (2016) The politics of evidence-based policymaking [online]. *The Guardian*, 10 March. Accessed 2 June 2019. Available at: https://theguardian.com/science/political-science/2016/mar/10/the-politics-of-evidence-based-policymaking

Cook, J. (2017). Response by Cook to 'Beyond counting climate consensus'. *Environmental Communication.* **11**(6): 733–735.

Cook, J. and Lewandowsky, S. (2016) Rational irrationality: modeling climate change belief polarization using Bayesian networks. *Topics in Cognitive Science.* **8**(1): 160–179.

Cook, J., Lewandowsky, S. and Ecker, U.K. (2017) Neutralizing misinformation through inoculation: exposing misleading argumentation techniques reduces their influence. *PLOS ONE.* **12**(5): e0175799.

Cook, J., Nuccitelli, D., Green, S.A., Richardson, M., Winkler, B., Painting, R., Way, R., Jacobs, P. and Skuce, A. (2013) Quantifying the consensus on anthropogenic global warming in the scientific literature. *Environmental Research Letters.* **8**(2): 024024+.

Cook, J., Oreskes, N., Doran, P.T., Anderegg, W.R., Verheggen, B., Maibach, E.W., Carlton, J.S., Lewandowsky, S., Skuce, A.G., Green, S.A. and Nuccitelli, D. (2016) Consensus on consensus: a synthesis of consensus estimates on human-caused global warming. *Environmental Research Letters.* **11**(4): 048002.

Cook, J., van der Linden, S., Maibach, E. and Lewandowsky, S. (2018). *The Consensus Handbook: Why the Consensus on Climate Change is Important.* Fairfax, VA: George Mason University. doi:10.13021/G8MM6P.

Deryugina, T. and Shurchkov, O. (2016) The effect of information provision on public consensus about climate change. *PLOS ONE.* **11**(4): e0151469.

Ding, D., Maibach, E.W., Zhao, X., Roser-Renouf, C. and Leiserowitz, A. (2011) Support for climate policy and societal action are linked to perceptions about scientific agreement. *Nature Climate Change.* **1**(9): 462–466.

Dixon, G., Hmielowski, J. and Ma, Y. (2017) Improving climate change acceptance among US conservatives through value-based message targeting. *Science Communication*, **39**(4), 520–534.

Doran, P.T. and Zimmerman, M.K. (2009) Examining the scientific consensus on climate change. *Eos, Transactions American Geophysical Union.* **90**(3): 22–23.

Dunwoody, S. and Kohl, P.A. (2017) Using weight-of-experts messaging to communicate accurately about contested science. *Science Communication.* **39**(3): 338–357.

Elsasser, S.W. and Dunlap, R.E. (2013) Leading voices in the denier choir: conservative columnists' dismissal of global warming and denigration of climate science. *American Behavioral Scientist.* **57**: 754–776.

Goldberg, M.H., van der Linden, S., Ballew, M.T., Rosenthal, S.A. and Leiserowitz, A. (2019) The role of anchoring in judgments about expert consensus. *Journal of Applied Social Psychology.* **49**(3): 192–200.

Hamilton, L.C. (2016) Public awareness of the scientific consensus on climate. *SAGE Open.* **6**(4): 2158244016676296.

Harris, A.J., Sildmäe, O., Speekenbrink, M. and Hahn, U. (2019) The potential power of experience in communications of expert consensus levels. *Journal of Risk Research.* **22**(5): 593–609.

Hornsey, M.J., Harris, E.A., Bain, P.G. and Fielding, K.S. (2016) Meta-analyses of the determinants and outcomes of belief in climate change. *Nature Climate Change.* **6**(6): 622–626.

Jasanoff, S. (2010) A new climate for society. *Theory, Culture & Society.* **27**(2–3): 233–253.

Kahan, D.M. (2015) Climate-Science Communication and the Measurement Problem. *Political Psychology.* **36**(S1): 1–43.

Kahan, D.M. (2016) The 'gateway belief' illusion: reanalyzing the results of a scientific-consensus messaging study. *Journal of Science Communication.* **16**(5): 1–20.

Kahan, D.M. and Carpenter, K. (2017) Out of the lab and into the field. *Nature Climate Change.* **7**(5): 309.

Kahan, D.M., Jenkins-Smith, H. and Braman, D. (2011) Cultural cognition of scientific consensus. *Journal of Risk Research.* **14**(2): 147–174.

Kerr, J.R. and Wilson, M.S. (2018a) Perceptions of scientific consensus do not predict later beliefs about the reality of climate change: a test of the gateway belief model using cross-lagged panel analysis. *Journal of Environmental Psychology.* **59**: 107–110.

Kerr, J.R. and Wilson, M.S. (2018b) Changes in perceived scientific consensus shift beliefs about climate change and GM food safety. *PLOS ONE.* **13**(7): e0200295.

Kirilenko, A.P., Molodtsova, T. and Stepchenkova, S.O. (2015) People as sensors: mass media and local temperature influence climate change discussion on Twitter. Global Environmental Change. **30**(Supplement C): 92–100.

Kobayashi, K. (2018) The Impact of perceived scientific and social consensus on scientific beliefs. *Science Communication.* **40**(1): 63–88.

Lewandowsky, S., Gignac, G.E. and Vaughan, S. (2013) The pivotal role of perceived scientific consensus in acceptance of science. *Nature Climate Change.* **3**(4): 399–404.

Lewandowsky, S., Oreskes, N., Risbey, J.S., Newell, B.R. and Smithson, M. (2015) Seepage: climate change denial and its effect on the scientific community. *Global Environmental Change.* **33**: 1–13.

Luntz, F. (2003) The environment: a cleaner, safer, healthier America [online]. *Luntz Research Companies*, Washington DC. Accessed 12 June 2019. Available at: https://the-republican-reversal.com/uploads/1/2/0/2/120201024/luntzresearch.memo2.pdf

Ma, Y., Dixon, G. and Hmielowski, J.D. (2019) Psychological reactance from reading basic facts on climate change: the role of prior views and political identification. *Environmental Communication: A Journal of Nature and Culture.* **13**(1): 71–86.

Maibach, E.W. and van der Linden, S.L. (2016) The importance of assessing and communicating scientific consensus. *Environmental Research Letters.* **11**(9): 091003.

Malka, A., Krosnick, J.A. and Langer, G. (2009) The association of knowledge with concern about global warming: trusted information sources shape public thinking. *Risk Analysis: An International Journal.* **29**(5): 633–647.

McCright, A.M., Charters, M., Dentzman, K. and Dietz, T. (2016) Examining the effectiveness of climate change frames in the face of a climate change denial counter-frame. *Topics in Cognitive Science.* **8**(1): 76–97.

McCright, A.M., Dunlap, R.E. and Xiao, C. (2013) Perceived scientific agreement and support for government action on climate change in the USA. *Climatic Change.* **119**(2): 511–518.

Miller, C. A. (2004). Climate science and the making of a global political order. In S. Jasanoff, ed. *States of Knowledge: The Co-Production of Science and the Social Order.* London: Routledge. pp. 46–66.

Myers, T.A., Maibach, E., Peters, E. and Leiserowitz, A. (2015) Simple messages help set the record straight about scientific agreement on human-caused climate change: the results of two experiments. *PLOS ONE.* **10**(3): e0120985-e0120985.

Oreskes, N. (2004) The scientific consensus on climate change. *Science.* **306**(5702): 1686.

Oreskes, N. and Conway, E.M. (2010) *Merchants of Doubt: How a Handful of Scientists Obscured the Truth on Issues from Tobacco Smoke to Global Warming.* New York: Bloomsbury Publishing.

Pearce, W., Brown, B., Nerlich, B. and Koteyko, N. (2015) Communicating climate change: conduits, content, and consensus. *WIREs Climate Change.* **6**(6): 613–626.

Pearce, W., Grundmann, R., Hulme, M., Raman, S., Hadley Kershaw, E. and Tsouvalis, J. (2017a) Beyond counting climate consensus. *Environmental Communication.* **11**(6): 723–730.

Pearce, W., Grundmann, R., Hulme, M., Raman, S., Hadley Kershaw, E. and Tsouvalis, J. (2017b) A reply to Cook and Oreskes on climate science consensus messaging. *Environmental Communication.* **11**(6): 736–739.

Powell, J.L. (2011) *The Inquisition of Climate Science.* New York: Columbia University Press.

Ranney, M.A. and Clark, D. (2016) Climate change conceptual change: scientific information can transform attitudes. *Topics in Cognitive Science.* **8**(1), 49–75.

Readfearn, G. (2013) The campaigns that tried to break the climate science consensus [online]. *Desmogblog.* Accessed. 2 June 2019. Available at: http://desmogblog.com/2013/06/06/campaigns-tried-break-climate-science-consensus

Readfearn, G. (2016) Revealed: most popular climate story on social media told half a million people the science was a hoax [online]. *Desmogblog..* Accessed 2 June 2019. Available at: https://desmogblog.com/2016/11/29/revealed-most-popular-climate-story-social-media-told-half-million-people-science-was-hoax

Rescher, N. (1995) *Pluralism: Against the Demand for Consensus.* Oxford: Clarendon Press.

Rolfe-Redding, J., Maibach, E.W., Feldman, L. and Leiserowitz, A. (2011) Republicans and climate change: An audience analysis of predictors for belief and policy preferences [online]. *SSRN 2026002.* Accessed 2 June 2019. Available at: http://papers.ssrn.com/abstract=2026002

Schuldt, J.P. and Pearson, A.R. (2016) The role of race and ethnicity in climate change polarization: evidence from a US national survey experiment. *Climatic Change.* **136**(3–4): 495–505.

Tom, J.C. (2018) Social origins of scientific deviance: examining creationism and global warming skepticism. *Sociological Perspectives.* **61**(3): 341–360.

Trench, B. (2008) Towards an analytical framework of science communication models. In D. Cheng, M. Claessens, T. Gascoigne, J. Metcalfe, B. Schiele, and S. Shi, eds. *Communicating Science in Social Contexts: New Models, New Practices.* Dordrecht: Springer. pp. 119–135.

van der Linden, S., Leiserowitz, A. and Maibach, E. (2018b) Perceptions of scientific consensus predict later beliefs about the reality of climate change using cross-lagged panel analysis: a response to Kerr and Wilson (2018). *Journal of Environmental Psychology.* **60**: 110–111.

van der Linden, S., Leiserowitz, A. and Maibach, E. (2019) The gateway belief model: A large-scale replication. *Journal of Environmental Psychology.* **62**: 49–58.

van der Linden, S., Leiserowitz, A., Rosenthal, S. and Maibach, E. (2017a) Inoculating the public against misinformation about climate change. *Global Challenges.* **1**(2): 1600008.

van der Linden, S., Maibach, E., Cook, J., Leiserowitz, A., Ranney, M., Lewandowsky, S., Árvai, J. and Weber, E.U. (2017b) Culture versus cognition is a false dilemma. *Nature Climate Change.* **7**(7): 457.

van der Linden, S.L., Clarke, C.E. and Maibach, E.W. (2015a) Highlighting consensus among medical scientists increases public support for vaccines: evidence from a randomized experiment. *BMC Public Health.* **15**(1): 1207.

van der Linden, S.L., Leiserowitz, A.A., Feinberg, G.D. and Maibach, E.W. (2014) How to communicate the scientific consensus on climate change: plain facts, pie charts or metaphors? *Climatic Change.* **126**(1–2): 255–262.

van der Linden, S.L., Leiserowitz, A.A., Feinberg, G.D. and Maibach, E.W. (2015b) The scientific consensus on climate change as a gateway belief: experimental evidence. *PLOS ONE.* **10**(2) e0118489.

Watts, J. (2018) We have 12 years to limit climate change catastrophe, warns UN [online]. *The Guardian* 8 October. Accessed 2 June 2019. Available at: https://theguardian.com/environment/2018/oct/08/global-warming-must-not-exceed-15c-warns-landmark-un-report

Weiss, C.H. (1979) The many meanings of research utilization. *Public Administration Review.* **39**(5): 426–431.

Whitmarsh, L. and Corner, A. (2017) Tools for a new climate conversation: a mixed-methods study of language for public engagement across the political spectrum. *Global Environmental Change.* **42**: 122–135.

Wilsdon, J. and Willis, R. (2004) *See-Through Science: Why Public Engagement Needs to Move Upstream*. London: Demos.

Worth, K. (2017) Climate change skeptic group seeks to influence 200,000 teachers [online]. *PBS Frontline...* Accessed 2 June 2019. Available at: https://pbs.org/wgbh/frontline/article/climate-change-skeptic-group-seeks-to-influence-200000-teachers/

Zhang, B., van der Linden, S., Mildenberger, M., Marlon, J.R., Howe, P.D. and Leiserowitz, A. (2018) Experimental effects of climate messages vary geographically. *Nature Climate Change.* **8**(5): 370.

10 Do rich people rather than rich countries bear the greatest responsibility for climate change?

Paul G. Harris and Kenneth Shockley

Summary of the debate

This debate engages with the question of moral responsibility and climate change. We might agree that the rich carry more responsibility for causing the climate to change than do the poor, but is it rich *people* or rich *countries* that should be held to greater account? **Paul Harris** argues that given the huge inequalities *within* states—both developed and developing—it should be rich individuals who carry the greatest moral responsibility for tackling climate change, although rich states have enormous responsibilities, too. People's actions cause greenhouse gases to be emitted and the more one emits the greater ones responsibility. **Kenneth Shockley** challenges this position by arguing that it is nation states that are the appropriate, and more effective, agents for acting to reduce the risks of climate change. Individuals, especially rich individuals, do have some moral responsibility, but primarily it is states who must accept the greater responsibility for climate change and lead the search for solutions.

YES: Rich people ought to behave responsibly (before it's too late) (*Paul G. Harris*)

Introduction

Contrary to received wisdom, countries—that is, nation states—do not cause climate change; people do. People are the agents who actually behave in ways that cause climate change. If every country that currently exists were to be dissolved, climate change would continue to grow worse because the agents of change—people—would continue to behave in ways that directly and indirectly emit greenhouse gases (GHGs) that pollute the atmosphere. In contrast, if every citizen of every country were to disappear, climate change would be addressed to the fullest extent possible.

The most basic conceptions of fairness and justice, which even children understand, demand that those who are responsible for causing harm should stop doing so. Those who cause the most harm, especially if they do so voluntarily and for relatively trivial reasons, have the greatest responsibility to stop. With few exceptions, the richer the person the more that person consumes and pollutes and thus the greater that person's adverse environmental impact. What is more, the richer the person the more that person is capable of choosing to behave differently and the more capable that person is to help others to do likewise—and indeed to aid those who suffer the consequences of pollution. Consequently, *rich* people have disproportionately greater responsibility for climate change and thus disproportionately greater responsibility to stop making the problem worse. As Simon Caney has argued, 'the burden of dealing with climate change should rest predominantly with the wealthy of the world, by which I mean affluent persons in the world (not affluent countries)' (Caney, 2005: 770).

With riches come responsibilities

What does it mean to be a *rich* person? To be rich is to have 'abundant possessions and material wealth', with 'more than enough to gratify normal needs' (*Merriam-Webster Dictionary*). By this definition, there are well over one billion rich people around the world. The majority of them live in Europe, North America and other parts of the 'developed' world, but at least hundreds of millions of them live within 'developing' countries (such as Brazil, China and India). Most readers of this book are likely to be rich. Even if you do not have a large amount of money in the bank, if you have more than enough 'stuff' to gratify your needs, you are rich, especially when compared to how billions of people continue to live (more than a few of them in rich countries). Cheeseburgers, fast fashion, luxury possessions, energy-hungry homes, private automobiles, cheap air travel and many other manifestations of affluent 'western' lifestyles are, historically speaking, not normal. Neither are they normal if there is any hope of avoiding catastrophic climate change.

To put it melodramatically, if every person who is rich by this definition were to drop dead tomorrow, the nominal objectives of the **Paris Agreement on Climate Change** for the next few decades would probably be realised without any additional effort by countries, rich or poor (cf. Chakravarty et al., 2009). Rich people do matter. Naturally, measured individually, super-rich people matter more than ordinarily rich ones; each rich person is responsible for contributing to climate change and each super-rich person is even more responsible (Otto et al., 2019). Nevertheless, nearly all rich people live and consume like never before in human history (and often have one or more children who will do the same throughout *their* lives). They live well beyond meeting their needs, polluting the atmosphere in the process. Vitally, the number of people living in this way is growing very rapidly around the world.

Without discounting the enormous responsibility of rich *countries* for climate change, putting all of the responsibility on them is a recipe for dangerous climate change. Even if one believes that rich countries deserve to be blamed (as I do), and even if those countries accept this blame (many of them claim that they do), very few of them have acted on this responsibility to anywhere near the extent that is required to slow global warming and other manifestations of climate change. We have far too little to show for blaming rich countries for climate change. Doing so almost exclusively is neither a practical way of attributing responsibility nor the most morally correct way to do so.

Waving the flag of responsibility: blaming rich countries

Before we can settle on the fairest solutions to the problem, and specifically who or what ought to take which actions, we need to decide who or what is *most* responsible. International negotiations and national policy formulation around climate change have been dominated by the mantra that rich countries ought to be taking responsibility. This idea is at the root of the concept of 'common but differentiated responsibility' that underlies international negotiations on climate change: while all countries share responsibility, the rich ones deserve most of it, and they therefore ought to be taking action to cut their climate-changing emissions and to help developing countries cope with the consequences of those emissions.

This initial focus on countries as the responsible agents was not irrational. Nearly all other global problems—war and peace, economic development and even most trans-boundary environmental problems—have been dealt with by countries trying to act collectively. They have often done so successfully. What is more, although states are formally only institutional constructs (a state per se does not have any agency; see **Chapter 12**), most people identify as nationals of a particular country, they need passports to travel very far internationally and there are physical barriers between many countries. But climate change does not recognise political boundaries. Neither the power of the state nor passport controls nor border walls nor anything else yet devised can stop GHG emissions 'going global' and affecting almost every community. No country has yet found a way to immunise itself against the effects of climate change. Such a solution does not exist (although for a counter-argument see **Chapter 8**).

As rich and poor countries trade accusations in international negotiations, and national leaders point to other countries' inaction to combat climate change as justification for their own lack of action (this is precisely the argument used by governments of Australia and the USA), GHG emissions continue to increase. This is happening even as scientists tell us that we need to all but eliminate them very soon (Xu et al., 2018). Despite several decades of negotiations by countries to solve this problem—starting before most of the readers of this book were born—GHG emissions into the atmosphere *continue to rise*. It is patently obvious that countries have failed to address climate change effectively and there is no indication that this will change anytime soon. Because the world's responses to climate change have focused on the rights and responsibilities of *countries*, resulting international agreements and related national policies have largely ignored the rights and responsibilities of *people* (see Harris, 2016). Aggressive responses to climate change cannot come from continuing on this pathway. Without major and widespread action by rich people very soon, it seems all but inevitable that climate change will be catastrophic, especially for the world's poor (see **Chapter 1**).

In addition to the unwillingness of rich countries to act responsibly, a problem with the country-focused attribution of responsibility is that it has let rich people largely off the hook, in both practical and moral terms. Blaming rich countries for climate change has made rich people everywhere lazy. They are waiting for rich country governments to do all of the work of fostering action. Rich people can say to themselves, 'I have paid my taxes and done what my government has demanded of me. I have fulfilled my responsibility'. But this attitude is wrong both practically and morally: it enables people who contribute disproportionately to the problem to pretend that they are not disproportionately responsible to change their related behaviours. It allows them to imagine that the moral duties fall on other actors. And it allows those rich people who live in developing countries—a smaller number than in developed countries, but one that is growing rapidly—to ignore their responsibilities completely. Is it any wonder that there is so little willingness among citizens to embrace the major changes in their lives that are needed to face climate change squarely?

A cosmopolitan debate: responsibilities of the most capable people

There is an alternative to the assumption that climate change is primarily the responsibility of rich countries. That alternative can be found in **cosmopolitanism**, the notion that all human beings belong to a single community, based on a shared morality. Cosmopolitanism attributes responsibility (and rights) to persons (as well as other actors). From a cosmopolitan perspective, people are moral agents who should take responsibility for their actions. Cosmopolitans acknowledge the responsibilities of capable people regardless of the countries in which they reside or hold citizenship. While cosmopolitans would not deny the responsibilities of rich countries, neither would they deny the responsibilities of rich people (Harris, 2016). For cosmopolitans, a rich person in Britain is not, *prima facie*, any more (or less) responsible for climate change than is a rich person in China. From this perspective, rich people ought not wait for rich countries to act on those countries' responsibilities for climate change. Rich people are responsible to do what they can to mitigate their own contributions to climate change regardless of what rich countries do.

Cosmopolitanism is a useful way to debate climate change. It exposes the reality that many millions of rich people around the world remain under little or no legal or even moral obligation to do anything about climate change solely because they do not live within the rich countries that have started to regulate climate damaging behaviours (see Harris, 2016). To blame rich countries for climate change is to *not* blame millions of the richest people on Earth. For example, from a country-focused perspective, a rich German enjoying a flight in her private jet is held *indirectly* responsible for climate change (but only a little bit, because she is still enjoying the flight after all) because her behaviour is taxed by her country's government as part of its efforts to reduce the country's (and, technically, also the EU's) GHG emissions. However, an even richer Brazilian enjoying a flight in his own private jet bears no responsibility at all, not even indirectly, because his country is not technically 'rich'.

Similarly, a well-off (but not super-rich) resident of rich Sweden will have his behaviour taxed and regulated to push him to reduce his activities that lead to GHG emissions. In contrast, millions of better-off people (vastly more than the entire population of Sweden) in 'developing' China are encouraged by their government to consume and travel to fuel economic growth—to do all of the things in fact that people must *stop* doing if climate change is to be taken seriously. Cosmopolitanism reveals this absurdity by looking past the political importance of national borders to identify where responsibility ought to be attributed. In doing so, it reveals both the ethical and practical importance of making *rich people everywhere* responsible for climate change. Even if not all climate policy revolves around what is revealed by this alternative approach, much of it ought to be.

All rich people, regardless of whether they live in rich countries, ought not wait for their national governments to force them to change their polluting lifestyles. They should reduce their GHG emissions as far as they can, which effectively means trying to eliminate all non-essential polluting activities from their lives. For example, rich people ought proactively to curtail airline travel because such behaviour contributes substantially to climate change and is almost always not necessary, often done only for pleasure. Rich people ought to curtail their consumption of meat because it is a major source of GHG emissions, is not necessary for health and alternative foods are available to almost all rich people. Rich people ought to have fewer children—unless those children will be nurtured to live their lives sustainably.

Rich people should also use their capabilities to push for public policies that foster major cuts in GHG emissions. Rich people in democracies ought to support and vote for candidates who advocate aggressive pro-climate policies. Rich people ought to take such actions not because their countries are responsible for climate change, but because *they* are as well. To emphasise this point, James Garvey asks us to look in the mirror: 'It is possible to think that my failure to do something about my high-carbon lifestyle really is morally outrageous' (Garvey, 2008: 142; cf. Harris, 2016: 151–152).

What individual persons do, and of course especially what individual rich persons do, matters because each individual's contribution to climate-changing pollution is added to that of everyone else's. As Steve Vanderheiden argues, 'Isolated individual contributions to larger aggregate problems may appear to be trivial, yet the countless occurrences of such seemingly trivial acts together add up to quite serious harms' (Vanderheiden, 2008: 166). To use the words of Thomas Pogge, which he applies to the problem of global poverty: 'nearly every privileged person might say that she bears no responsibility at all because she alone is powerless to bring about a reform of the global order' (Pogge, 2002: 170). Pogge describes this as 'an implausible line of argument, entailing as it does that each participant in a massacre is innocent, provided any persons killed would have been killed by others, had he abstained' (Pogge, 2002: 170).

Many will argue that individual rich persons actually do not have power to respond to climate change in meaningful ways. Garvey has a response to such thinking:

> against the claim that individual choices cannot matter much, is that nothing else about you stands a chance of making a moral difference at all. If anything matters, it's all those little choices … The only chance you have of making a moral difference consists in the individual choices you make.
>
> (Garvey, 2008: 150)

Put another way,

> the total impact of a life lived high on the hog compared to one lived simply adds up and, when multiplied by two billion or more other relatively affluent people in the world, the impact is gargantuan. It is the difference between a liveable planet for all and truly monumental suffering for billions.
>
> (Harris, 2016: 192–193)

Thinking more about what rich people do, and less about what rich countries do, helps to highlight these realities.

Historical responsibility: the practicality of a paradox

Looking at responsibility for climate change from a cosmopolitan perspective has other benefits. For example, it enables us to do a better job of assessing historical responsibility for climate change. One obstacle to persuading rich countries to take on more responsibility, and to act accordingly, is differing views on which of them are more or less responsible for GHG emissions in the past. By looking at responsibility from a cosmopolitan perspective, one might see that there could be more responsibility among rich people in *developing* countries than normal country-oriented arguments might reveal. This is because rich people in affluent countries did not realise that they were causing climate change until the latter years of the last century. When they realised the consequences of what they were doing, they were already deeply immersed in lifestyles that lead to severe climate change.

In contrast, rich people in many developing countries who adopted similar lifestyles only recently, including the many millions of new middle-class consumers in those countries, knew from the time that they became 'rich' that many of their consumption behaviours were contributing to climate change. They were aware of climate change before getting on the global consumption bandwagon and with that awareness comes responsibility. History may not judge them favourably. (We can debate whether ordinary citizens fully understand their own contribution to climate change, but we ought not overlook wilful ignorance among most rich people.)

Considering such a paradoxical conclusion is only possible if we stop focusing so much on rich countries and instead focus more on rich people, including those who do not live in rich countries. For example, the number of Chinese tourists is growing rapidly, helping to make air travel one of the fastest-growing sources of climate-changing emissions. Millions of Chinese citizens now fly around the world on holidays, contributing greatly to climate change. If we attribute responsibility for climate change to rich *countries*, all of these holidaymakers bear no responsibility for the contribution that they make to climate change because China (despite its obvious wealth) is still officially 'developing'. Those Chinese tourists can and do make the argument that they have a right to consume in this way because China does not bear responsibility and that it has the right to follow the same development path as that of the world's rich countries (see Harris, 2011).

Debating climate change from a cosmopolitan perspective could lead to better policies for and by countries. In addition to doing what is morally right, assuming more responsibility for climate change by rich people—even those not living in rich countries—could help to nudge rich countries to do more to live up to *their* responsibilities for climate change. For example, if rich people outside rich countries are not held responsible for climate change, rich people within rich countries—most of which are democracies, where rich people have significant influence on policymaking—will do all that they can to discourage their governments from implementing policies that hold the rich sufficiently responsible. This is more or less what has been happening for decades, with the consequence being very few new policies that result in major cuts in people's GHG emissions. (Has any rich person in any rich country avoided air travel due to a government policy discouraging such behaviour?) It would be much harder for rich people in rich countries to justify such responses to climate change if rich people all around the world were made proportionately responsible. What is more, if not-so-rich people in rich countries were to see the richest among them finally taking responsibility and acting accordingly, widespread political opposition to major action on climate change would be diminished.

Conclusion

We can argue that rich people have the greatest responsibility for climate change because they cause so much harm and are so capable of stopping that harm. But, in truth, they are not solely responsible. Rich countries, too, are responsible—as are all capable actors, ranging from community groups and businesses to multinational corporations and sub-state actors, such as Scotland and New York City, as well as international governmental and nongovernmental organisations, such as the European Union and the Fédération Internationale de Football Association (FIFA). Collectively, all of them are responsible for climate change (see **Chapter 12**). Collectively, all actors that are rich have a responsibility to change their ways.

However, after three decades of effort, we should stop relying so heavily on countries to do that which they are not well designed to do: to act for the long-term global collective good instead of for their own perceived short-term national interests (which, by the way, are often shaped to protect the interests of their most affluential citizens). While we ought not ignore the responsibilities of rich countries, we also ought not ignore the responsibilities of rich people. More of the climate change debate should be about them. More policies should be about them. More education should be about their responsibilities—about your responsibilities, and mine.

And we should debate the fact that millions of the world's rich people live in developing countries. Regardless of what those countries are willing and able to do to address climate change, their rich citizens ought not be let off the hook. To do so sends the worst message to rich people in rich countries. Telling the latter to bear fewer children, fly less, drive less, eat less meat, live in smaller homes, use less energy and consume less, will not be nearly as effective a message if they see rich people in developing countries flying more, eating more meat and so on. This goes without saying, but we ought to start saying it much more. Again, rich people matter—everywhere.

Acknowledgements

The author has made similar arguments in Harris (2013, 2016) and in other publications, which are listed on his website: www.paulgharris.net.

NO: *Primary* responsibility must rest with states and institutional actors (*Kenneth Shockley*)

Introduction: what is moral responsibility?

Rich people do not bear the greatest responsibility for climate change. *Primary* responsibility falls on states, provincial and other regional governments, international corporations, financial institutions, intergovernmental organisations and other actors in the international arena. In what follows I will refer to these entities as institutional actors, in contrast to individual actors. Institutional actors are able to bring about change, are the primary drivers for the climate change that has taken place and are the primary means by which it can be addressed. Whether we think of responsibility as retrospective (the responsibility we hold for what we have done historically) or prospective (the responsibility we hold for what we can do), the responsible party must be a causally *effective* agent. And in the international arena where responsibility for climate change can be meaningfully addressed, individuals are simply not effective agents. The argument is straightforward: if something is not an effective agent, then it cannot be morally responsible. First, rich people, collectively, do not constitute an agent, and so do not constitute an effective agent. Second, a rich person as an individual does not have sufficient causal control over climate change and so does not constitute an effective agent, at least in relation to climate change. As moral responsibility requires some degree of causal control on the part of a responsible agent, rich people are not morally responsible for climate change … at least not directly.

We should be clear on what is necessary to be held responsible.[1] Whether a matter of being held responsible for what we have done in the past or what we are required to do going forward, moral responsibility requires both causal connection and blameworthiness. The first element, causal connection, is the requirement that to be retrospectively responsible for something one must have *done* it, and to be prospectively responsible for something one must *be able* to do it. On most accounts of responsibility there must be some causal connection between an agent's performance of an action and any responsibility attributed to that agent for the action or state of affairs that results from that action. For example, suppose an anvil falls off a building and hits Abbas on the head. If Beta neither caused the anvil to fall (say, by pushing it), nor could have prevented the anvil from falling (say, because she lacked the strength), it would be odd to say that Beta is either retrospectively or prospectively morally responsible for what happened to Abbas. Moral responsibility requires a degree of causal control sufficient to make an individual an effective agent. Unless an individual is causally effective, they cannot be morally responsible.

The second element, blameworthiness, amounts to the claim that there was something culpable about what was done (e.g. it was done recklessly) or the way it was done (e.g. it was done with malicious intent), or what should be done (as a duty of one's social station or as compensation for a historical benefit or historical transgression). But note, there is an important asymmetry between causal control and blameworthiness. In order to be blameworthy for an action or outcome one must have some causal connection to the action or outcome. It isn't appropriate to blame someone for an action, to hold them responsible for it, when they neither performed that action (they didn't *do* it) nor could have prevented it. Causation is required before we can consider blame. For example, even if someone steps on my toe (and so has the requisite causal connection), they are only blameworthy for stepping on my toe if they did it intentionally, or if it occurred because they were negligent and failed to exercise due care, or could and should have taken measures to avoid stepping on my toe. The acid test for blameworthiness is whether or not the relevant party had *mens rea*, a guilty mind. We will return to blameworthiness in the discussion below but, here, we should see the central importance of causal control for responsibility.

Rich people, responsibility and effective agency

While in aggregate rich people have had a disproportionate influence on the climate, they are individually not effective agents and, in aggregate, not agents at all. First, once we are clear about the significance of causal control for matters of moral responsibility, it should be equally clear that individual agents are not individually responsible for climate change (Sinnott-Armstrong, 2005). It is certainly the case that a very small percentage of the world's population engage in activities that produce a disproportionate share of emissions (Shue, 1993). With very few exceptions, however, only states and other international actors have the requisite causal control over emissions and so only states and other international actors are, therefore, responsible. Particular rich individuals are generally not *effective* agents of change either for addressing future challenges generated by climate change or for taking historical responsibility for climate change. Moreover, exceptions help make the case. Some very few wealthy or politically powerful individuals do have what we might think of as international agency, including the Pope, Bill Gates, Jeff Bezos and others. But they usually have this agency only through institutional actors (the Church, international corporations and so on). This connection between individuals and institutions will be important in what follows.

Second, rich people are not collectively responsible for climate change as they are not the sort of group that is or could be a collective agent. We should be clear on what would be required for 'rich people' to be an agent. Following List and Pettit (2011: 158), a group is morally responsible if they are able, *as a group*, (1) to face a 'normatively significant choice' that involves performing an action that is right or wrong, or good or bad, (2) to understand that choice in such a way that they can evaluate options and make a 'normative judgement about the options,' and (3) to have 'the control required for choosing between the options'. The key point is that for the group to be held responsible it will have to have the capacity for deliberation and control necessary for any agent. Rich people, taken as a single group, are simply not the right sort of entity. All rich individuals, taken together, do not share the deliberative or decision-making processes or institutions necessary for attributing agency. There is no collective understanding, evaluation, judgement or control. There is no capacity for collective *choice*. Therefore, as a group, rich people do not constitute an agent of the relevant sort to hold either backward directed responsibility for historical harms or forward directed responsibility for making the necessary changes.

But this doesn't get rich people off the hook (see Moellendorf, 2014). Individuals, particularly rich individuals, are *not* morally free to do whatever they like, thereby offloading moral responsibility to institutional actors. They have responsibilities, but those responsibilities are filtered, indirectly, through states and other institutional actors.

States and institutions are the appropriate focus for responsibility

In contrast with rich individuals, institutional actors *are* effective agents in the international arena (see List and Pettit, 2011, for an argument that institutional entities may constitute agents). States are able to affect domestic policy by, for example, addressing consumption patterns and enabling individual actors to have effective choices involving greenhouse gas emissions. They are able to affect the international political context in a way conducive to addressing climate change. Other non-state institutional actors—banks, international political entities, provincial and other regional governments—also have the capacity to make similar changes in both the domestic and international political landscape (see **Chapter 12**). Decisions made by banks to support renewable energy resources, long-term infrastructure decisions made by local and regional governments and regional carbon trading markets (see **Chapter 6**), for example, all have the capacity to make substantial changes to emissions patterns.

Decisions by particular rich individuals usually have little or no effect on such patterns, except insofar as they leverage institutions. Not only are institutional actors (and remember, institutional actors include much more than traditional nation states) the relevant actors for addressing global environmental problems such as climate change, they are also the only means by which any particular individual could have any relevant effect in the international political realm.

Institutional actors are the mediators between individual choice and global change. And they are political actors. They can make a difference in how we address climate change; they can change the political landscape, both domestic and international, in which serious efforts to address climate change take place. Moreover, they are the entities that have failed to make the necessary changes, historically, that have put us in the place we are in. Primary responsibility for climate change rests with institutional actors. They hold and have held the levers to make the needed changes.

There is this important complication. As the concern is often about the *moral* responsibility associated with climate change, individuals might yet play a particularly significant role. We have seen that moral responsibility requires that the responsible entity have something akin to a guilty mind. This is challenging to see in the case of states and other institutional actors (although, again, see List and Pettit, 2011). Further, while it is unclear how rich people, taken as a group, could possibly have a guilty mind, it is quite clear that *individual persons*, rich or not, are more than capable of having a guilty mind.

Individuals are the ultimate example of moral agents. However, as we have noted, individuals are not causally connected to climate change in the right way and so do not bear (individual) moral responsibility for climate change (Jamieson, 1992). But individuals might well bear moral responsibility for failing to contribute to institutions in the right way (Gardiner, 2017; Shockley, 2016). For that is a matter over which they do have some control and a matter which may well have great moral significance.

What role should be played by the rich, or by individuals more generally?

Individuals do have a role to play, one that comes with substantial moral responsibility. In particular, rich individuals have benefitted the most from our unsustainable economic system, have had a disproportionate influence on the institutions that can have an effect on climate change and there are some who have a large amount of political influence. Rich individuals are therefore responsible for *contributing* to changes in institutional actors such that those institutional actors are better able to address climate change. With the great power that (rightly or wrongly) comes from benefitting from the economic systems that generate climate change comes great responsibility.

The industrial processes and economic growth that have fuelled climate change have made the rich rich. And so one might think there is a *prima facie* responsibility for providing some sort of compensation for the climate change associated with this wealth. However, focusing on this wealth may well lead us to miss the underlying cause—the institutional arrangements and economic and political systems—that enabled the development of individual wealth in the first place. If there is a responsibility on wealthy individuals for changing these arrangements and systems, then the greater the ability to make such changes, the greater the responsibility for doing so. The historical acquisition of power by rich individuals generates a responsibility to rectify wrongs—by changing the system that led to those wrongs. It also places a responsibility on them for making the forward directed changes that address the problems that will arise from climate change and to make the institutional changes necessary to minimise the generation of further harm. Because wealth provides the means to make the institutional changes that can have an effect on climate change, wealthy individuals have a responsibility to do so.

Yet one might worry that individuals will not live up to their responsibilities and make the necessary changes in their political and institutional environment. Getting individuals to live up to their responsibilities is clearly a problem, but it is a familiar one. The alternative, requiring individuals to satisfy direct cosmopolitan responsibilities for climate change—as argued here by **Paul Harris**—would seem to face a substantially worse version of this challenge. As Jamieson (1992) and Sinnott-Armstrong (2005) both make clear, it is very hard to operationalise, or even to make sense of, an individual responsibility for climate change, particularly a moral responsibility. In such contexts it is very difficult to see how any individual could be both blameworthy and causally connected to any relevant harm. Individuals may have difficulties living up to their well-established responsibilities. But it is much harder to see how individuals would live up to new and unfamiliar responsibilities they can't even understand. There is more hope in getting individuals to live up to their responsibilities through a contributory approach.

So, while the possession of economic and political power by rich people does bring with it some responsibility to drive institutional actors to change their behaviour, that responsibility is not for climate change directly. At least in most instances, individual responsibility is for making the changes (and having failed to make the changes in the past) to our political systems and the related economic and social institutions. We might well think of this form of individual moral responsibility as *contributory responsibility*. We will see below that this is important if we are to assess responsibility properly and bring about effective change: *both* institutions *and* individuals have roles to play.

Conclusion

There is a practical tension underlying the question of *who* bears primary responsibility for climate change. If primary responsibility is focused on rich individuals, we risk letting states off the hook. If states are not responsible, and rich individuals are, then there is no point in pressing states to address climate change. This echoes a common concern with the renewed focus on **non-state actors** in the climate change policy process (see **Chapter 12**)—provincial and regional governments, banks, nongovernmental organisations, research entities and so on. Some worry that states might reason that if *these* actors can play a larger role, then perhaps states can take a smaller role. They might thereby avoid having to address any historical responsibility or duties due to their citizens or to those affected by their historical actions. On the other hand, if primary responsibility is focused on states, we risk letting individuals, especially rich individuals, off the hook from making any individual changes. Individuals, it might be thought, can do whatever they want and pass the buck on to the institutions and governments that hold responsibility.

Of course, given the history of state inaction with regard to climate change, cynicism about states' ability to address climate change is understandable. Harris (2008: 482) writes, 'the climate change regime has failed. The arguments for international—that is, inter*state*—justice that have permeated the climate change regime have been insufficient'. His thought is that while we should still rely on states, we need to rely on individuals as well.

> To be sure, policy institutions (normally states) ought to play a big part by mediating the obligations of individual persons. However, institutions have failed so far; climate change is accelerating. We ought not reject the argument, made by some cosmopolitans, that people ought to push for the creation of the institutions that can mediate our obligations (Moellendorf, 2002). But we must be realistic in admitting the difficulty of doing this: we have not succeeded in doing it so far, we cannot wait forever, and huge numbers of people live in authoritarian environments where they have little ability to shape institutions, although they do often have the ability to shape their own behaviour.
>
> (Harris, 2008: 490)

Yet to think that individuals, acting through their reduced consumption, can affect change in the climate is at least as problematic as thinking states will solve all our climate problems for us. What I propose here is a middle ground: states and institutions bear *primary* responsibility, but individuals bear a *contributory* responsibility, a responsibility to ensure that those states and institutional actors are doing the job they are morally required to do (Shockley, 2016).

One of the great challenges of climate change is the dangerous decoupling of moral and causal responsibility, of both forward and backward directed varieties. Focusing on either individual responsibility, in order to capture the distinctively moral features of responsibility, or institutional responsibility, in order to capture the need for efficacious responsibilities, misses something. Contributory responsibility carries a distinctively moral connection between individual responsibility and the institutions that are effective climate actors in the international arena. It therefore provides a means of binding individuals to the moral harms of climate change, without requiring institutional actors to have moral responsibilities they cannot have or requiring individuals to do more than they are able. While the rich do not hold primary responsibility for climate change, we all have a responsibility to create and change the institutional actors that *are* the effective, responsible parties in our global response to climate change. To the extent that the rich are more able to bring about change in those institutions, they have a greater responsibility to do so.

Individuals have a better chance of leveraging institutions—states, financial institutions, NGOs—to thereby make the changes necessary to lead to less climate-intensive consumption. This is the contrast: no one should doubt that individuals have moral responsibilities resulting from the grossly problematic actions of states. The question is what they have responsibilities *for*. As individuals are ineffective actors in the international arena where climate change can be affected, they cannot be responsible for making those changes. Ought implies can, after all. But coordinated individuals, individuals in concert, organised by institutions and governments that they themselves reform or develop, *are* able to make the changes necessary to reduce emissions. And, if they are successful, to also reform the institutions and social arrangements that make climate change so deeply problematic.

If we want to address the cause of climate change we should look not to individuals, whether rich or not. Rather, we should look to the states and other institutions that are both the effective actors and mediators of individual actions. Primary responsibility for addressing climate change rests with those entities, even if it is the moral responsibility of individuals to change those entities so that they better address the greatest moral challenge of our day.

Further reading

Broome, J. (2012) *Climate Matters: Ethics in a Warming World*. New York: WW Norton & Company.

In this book, Broome considers the moral dimensions of climate change, reasoning through what universal standards of goodness and justice require of us, both as citizens and as governments. His conclusions both challenge and enlighten. Eco-conscious readers hear they have a duty to offset all their carbon emissions, while policymakers are called upon to grapple with what, if anything, is owed to future generations.

Gardiner, S.M. and Weisbach, D.A. (2016) *Debating Climate Ethics*. Oxford: Oxford University Press.

This book presents arguments for and against the relevance of ethics to global climate policy. Gardiner argues that climate change is fundamentally an ethical issue, since it is an early instance of a distinctive challenge to ethical action. Ethical concerns are at the heart of many of the decisions that need to be made. By contrast, Weisbach argues that existing ethical theories are not well suited to addressing climate change.

Harris, P.G. (2016) *Global Ethics and Climate Change*. Edinburgh: Edinburgh University Press.

This book combines the science of climate change with ethical critique. Harris exposes the increasing intensity of dangerous trends—particularly growing global affluence, material consumption and pollution—and the intensifying moral dimensions of changes to the environment. A free learning guide is available at: www.edinburghuniversitypress.com/media/resources/Global_Ethics_and_Climate_Change_2nd_Edition_-_Learning_Guide.pdf

Hayward, T. (2012) Climate change and ethics. *Nature Climate Change.* **2**(12): 843–848.

Hayward shows how the greater part of debate about the ethics of climate change focuses on questions about who has what responsibility to bear the burdens of mitigating it or adapting to it. The connections between human rights and climate change are examined, as too are questions concerning justice in the present, our responsibilities to the future and the relation between individual and collective responsibilities.

Peeters, W., Smet, A.D., Diependaele, L., Sterckx, S., McNeal, R.H. and De Smet, A. (2015) *Climate Change and Individual Responsibility: Agency, Moral Disengagement and the Motivational Gap.* Basingstoke: Palgrave Macmillan.

This book discusses the agency and responsibility of individuals for climate change. It argues that these responsibilities are underemphasised, enabling individuals to maintain their consumptive lifestyles without having to accept moral responsibility for their luxury emissions.

Follow-up questions for use in student classes

1. If rich countries fail to act on their responsibilities for climate change, which actors should do so instead? Why and how should those actors do it and why have they not done so already?
2. Would the responsibility of rich countries for climate change be different if all of their rich citizens moved to developing countries and took up citizenship there?
3. What level of wealth constitutes being rich enough to have a responsibility for changing political institutions capable of addressing climate change?
4. How effective are rich persons as agents of change?
5. What is *your* responsibility for causing climate change? What is *your* responsibility for trying to stop it?

Note

1 See also Moellendorf (2014: 152–180) for an excellent treatment of responsibility in the context of climate change.

References

Caney, S. (2005) Cosmopolitan justice, responsibility, and climate change. *Leiden Journal of International Law.* **18**: 747–775.

Chakravarty, S., Chikkatur, A., de Coninck, H., Pacala, S., Socolow, R. and Tavoni, M. (2009) Sharing global CO_2 emission reductions among one billion high emitters. *Proceedings of the National Academy of Sciences.* **106**(29): 11884–11888.

Gardiner, S. (2017). Accepting collective responsibility for the future. *Journal of Practical Ethics.* **5**(1): 22–52.

Garvey, J. (2008) *The Ethics of Climate Change: Right and Wrong in a Warming World.* London: Continuum.

Harris, P.G. (2008) Climate change and global citizenship. *Law and Policy.* **30**(4): 481–501.

Harris, P.G., ed. (2011) *China's Responsibility for Climate Change: Ethics, Fairness and Environmental Policy.* Bristol: Bristol University Press/Policy Press.

Harris, P.G. (2013) *What's Wrong with Climate Politics and How to Fix It.* Cambridge: Polity.

Harris, P.G. (2016) *Global Ethics and Climate Change.* Edinburgh: Edinburgh University Press.

Jamieson, D. (1992) Ethics, public policy, and global warming. *Science, Technology, and Human Values.* **17**(2): 139–153.

List, C. and Pettit, P. (2011) *Group Agency: The Possibility, Design, and Status of Corporate Agents.* New York: Oxford University Press.

Moellendorf, D. (2002) *Cosmopolitan Justice.* Boulder, CO: Westview Press.

Moellendorf, D. (2014) *The Moral Challenge of Dangerous Climate Change.* New York: Cambridge University Press.

Otto, I.M., Kim, K.M., Dubrovksy, N. and Lucht, W. (2019) Shift the focus from the super-poor to the super-rich. *Nature Climate Change.* **9**(2): 82–84.

Pogge, T. (2002) Human rights and human responsibilities. In P. De Grieff and C. Cronin, eds. *Global Justice and Transnational Politics: Essays on the Moral and Political Challenges of Globalization.* Cambridge, MA: MIT Press. pp. 151–196.

Shockley, K. (2016). Individual and contributory responsibility for environmental harm. In A. Thompson and S. Gardiner, eds. *The Oxford Handbook of Environmental Ethics.* Oxford: Oxford University Press. pp. 265–275.

Shue, H. (1993) Subsistence emissions and luxury emissions. *Law and Policy.* **15**(1): 39–59.

Sinnott-Armstrong, W. (2005) It's not *my* fault: global warming and individual moral obligations. In W. Sinnott-Armstrong and R. Howarth, eds. *Perspectives on Climate Change: Science, Economics, Politics, Ethics.* Amsterdam: Elsevier. pp. 285–307.

Vanderheiden, S. (2008) *Atmospheric Justice: A Political Theory of Climate Change.* Oxford: Oxford University Press.

Xu, Y., Ramanathan, V. and Victor, D.G. (2018) Global warming will happen faster than we think. *Nature.* **564**: 30–32.

11 Is climate change a human rights violation?

Catriona McKinnon and Marie-Catherine Petersmann

Summary of the debate

This debate considers whether climate change is a human rights violation and introduces students to normative political theory and the international human rights law (HRL) framework. **Catriona McKinnon** argues that climate change is already damaging a morally fundamental subset of human rights and will continue to do so in the future. All persons, present and future, have basic rights to subsistence and security and this generates a general duty for all people to work to create and support rights-respecting institutions. **Marie-Catherine Petersmann** challenges this position by highlighting the difficulties of applying the HRL framework to climate change which, by its very nature, is a collective action problem that cannot easily be linked to one duty bearer and one victim. Instead, she draws upon new insights from environmental humanities to offer a different way of thinking about co-responsible agents and beneficiaries of a safe climate.

YES: Because it undermines the right to life, to subsistence and to health (*Catriona McKinnon*)

Introduction

The political recognition of human rights is one of humanity's major moral achievements. At their most powerful human rights secure conditions in which every person has an opportunity to flourish. Taking human rights seriously places limits on what states and corporations can do to individuals and what individuals can do to one another. Human rights also compel states in particular to provide basic goods for people in need. International **human rights law** (HRL) aims to codify these duties. Three of the most fundamental human rights relevant to thinking about climate change are: *the right to life*; *the right to subsistence*, which secures access to goods necessary for survival such as food, clean water and shelter; and *the right to health*, which gives protection against avoidable illnesses and access to health care. Beyond HRL, the concern to ensure conditions in which these basic rights are protected has been characterised by the United Nations in terms of human security. The 1994 UN Human Development Report presents human security as embodying a 'concern with human life and dignity' and is focused on the importance of 'safety from the constant threats of hunger, disease, crimes and repression' (UNDP, 1994: 3).

Basic rights are a subset of human rights. Basic rights are rights without which people are unable to enjoy the full range of their human rights (Shue, 1996). Climate change presents a large-scale, serious and accelerating threat to the security of basic rights. Because basic rights are a moral minimum for a life of human dignity they ought to be placed at the centre of policy responses to climate change. No policy response to climate change can be considered adequate unless it achieves human security. Basic rights are held by people presently living. Many scholars argue that the same rights will be held by all people in the future. The moral importance of basic rights compels all of us to do what we can to protect them.

I shall argue that what it means for this requirement to be met differs according to the situation of the individual, or group of individuals, in question. This means that it is legitimate for us to require states and other groups (such as corporations) that have done most to cause climate change, and who can best afford to take on these demands, to bear the greatest costs of mitigation and adaptation adequate to protect basic human rights now and into the future (see **Chapter 10**). In practice this means that rich states, global economic elites and fossil fuel multinational corporations must commit to action to achieve a zero-carbon global economy by 2050. They should do so in ways that do not thwart the economic and political development necessary to secure basic human rights for people in the world's least advantaged countries.

Threats to basic rights from climate change

For hundreds of years political philosophers have argued for the importance of rights (Campbell, 2019). Some thinkers have conceived of rights as natural or God-given. Others have seen a commitment to rights for all as the logical conclusion of seeing one-self as deserving of rights. And others have seen rights as necessary to protect distinctively human interests and needs. Despite their differences there are some key features of rights agreed upon by all theorists. First, that human rights are universal: all people have the same rights. Second, that rights are the foundation of human flourishing: political protection for rights provides each person with a platform on which to build a successful life, whatever that might mean to them. Third, human rights are a tool by which we can hold one another to account in the political and social lives we must share together: respect for one another's rights is a basic rule that makes social and political society fair and stable. Fourth, that one of the most fundamental rights we have is the right to life: perhaps the most serious harm that can be inflicted on a person is to be deprived of their life.

The right to life is a **basic human right**. It is a right without which no other right can be enjoyed. There is an array of such basic rights—for example, liberty, physical integrity, subsistence and health. It is the job of the state to secure these basic rights for all its members, supported by international institutions where appropriate. There is a large debate in political philosophy over whether state boundaries are justified at all if we take seriously the equal basic rights of all people. For example, **cosmopolitan** thinkers argue that national boundaries are morally irrelevant, in contrast to those who treat the nation state as a site of special duties between citizens (Armstrong, 2019). Climate change presents grave threats to the basic rights of people around the world. It endangers food security, increases the likelihood of certain vector-borne diseases, makes dangerous heatwaves and flooding more common and may displace millions of people. People living in poverty are most vulnerable. This includes people living in low-lying coastal areas and small island states, poor people in megacities and those in regions most prone to extreme weather events and drought.

It is important to notice that climate threats to basic rights extend beyond the lifetimes of everyone alive today. Facts about the rate of change in the climate system given different rates and concentrations of greenhouse gas (GHG) emissions mean that certain climate impacts would continue for hundreds of years even if we were to achieve zero carbon emissions today (see **Chapter 2**). Because progress towards zero carbon emissions is slower than this, and because global emissions are still rising, people for thousands—perhaps hundreds of thousands—of years hence will be impacted by cumulative emissions of GHGs in ways that could violate their basic human rights. The threats to basic rights created by climate change require people in the same time slice to find political solutions that deliver justice both to other members of their temporal cohort and to people they will never know in the perhaps distant future. This ethical duty is part of what makes the search for climate justice a 'perfect moral storm' (Gardiner, 2011).

Tackling climate change in a way that minimises the violation of the basic rights of present and future people, especially those most vulnerable, requires (as a minimum) aggressive mitigation to achieve a zero-carbon global economy as soon as possible (see **Chapter 7**). If the basic rights of all people are to be respected, this goal must be pursued in ways that do not stunt development in less developed countries (LDCs) (see **Chapter 1**). Without a commitment to mitigate in ways that enable poor people to escape poverty, the people of these countries will face a double threat of climate impacts which exacerbate their existing vulnerabilities and which trap them in a state of deep human insecurity. This outcome would be seriously unjust, now and into the future.

What do rights require in the face of climate change?

Basic rights correlate with the duties of all people at least to refrain from acting in ways that would violate these rights and sometimes also to act positively so as to ensure the rights are met. This poses a problem for thinking about how to achieve rights-respecting climate justice. Who is required to act in the face of climate change if we take basic rights seriously? And what sort of action is demanded by rights-based climate justice?

Accepting that every person has a duty to refrain from acting in ways that would violate any other person's basic rights does not get us very far in finding answers to these questions. No one person's emissions directly cause climate impacts that violate any other person's basic rights. Instead, and as a minimum, respect for the basic rights of people who will be worst and soonest impacted by climate change requires each of us to act in ways that will make the political, social and economic changes necessary for a zero carbon economy more likely. Depending on an individual's circumstances this could mean mounting a political protest, signalling the need for changes in consumption in rich countries by going vegan, taking fewer flights or merely voting for political parties that are committed to such changes.

Beyond the moral minimum owed by each of us to show respect for the basic rights of climate-vulnerable people, political philosophers have argued that some groups of people have a duty to shoulder more of the burdens of taking action on climate justice than others. Philosophers have argued that in virtue of how these groups of people are related in particular ways to climate change they have special duties to take rights-respecting action on climate change that go beyond the duty every person has not to violate the basic rights of any other person. Philosophers have captured these special duties by specifying principles which distribute the burdens of justice-promoting action on climate change in ways that fall most heavily on these groups. The principles are: the polluter-pays principle (PPP); the beneficiaries-pay principle (BPP); and the ability-to-pay principle (APP). Most often these principles are applied to countries (see **Chapter 10**). Although not beyond dispute this makes sense given that countries are parties to the **Paris Agreement** (see **Chapter 12**).

Special duties

The PPP—the polluter-pays principle—asserts that those who have done most to cause climate change ought to do most to prevent it from violating the basic rights of present and future people. At a minimum, countries with high historical emissions should aggressively reduce their emissions to zero as soon as possible. If this is not sufficient to reduce cumulative emissions to a level at which threats to basic rights are averted, then these countries have duties to pursue **negative emissions** in order to keep the carbon budget within the necessary limits. Most IPCC scenarios for securing a 2°C target assume negative emissions, for example as a result of geoengineering through the use of bioenergy with **carbon capture and storage** (see **Chapter 7**).

Despite its intuitive force the PPP can be challenged. Climate change today and into the future is caused by the accumulation of GHGs in the atmosphere. CO_2 has a particularly long atmospheric lifetime. This means that climate change today and into the future is in part caused by emissions in the past. This supports an objection to the PPP: to assign special duties to people today in countries with historically high emissions would be unfair to those people. Present people in historically high-emitting countries are not the cause of climate change and have no special duties to take rights-respecting action on it. Parallels are sometimes drawn between this debate and the debate about reparations for slavery.

This objection to the PPP is countered by the BPP—the beneficiaries-pay principle. The BPP claims that people in countries that have benefitted from early carbon-intensive industrialisation ought to do most to secure the basic rights of people vulnerable to climate impacts. In contrast to the PPP the basis of the duty on this view is non-causal. The BPP asserts people who enjoy the advantages of early industrialisation—for example, better health, increased longevity, higher standards of education, time for leisure—have a special duty to ensure that people in LDCs can do the same. This commits them both to averting climate impacts that would violate the basic rights of these people and to creating opportunities for these countries to develop in ways that are consistent with keeping global warming below 2°C.

The BPP is vulnerable to the following criticism. Receiving a benefit creates a duty for a person if and only if that benefit was received willingly. Willing receipt of a benefit implies control over whether or not to receive it. In other words, if a benefit could not have been rejected it cannot be subject to willing receipt and so cannot serve as the basis of a duty on the part of the beneficiary. The objection is that this is exactly the situation of people in countries that have benefitted from early carbon-intensive industrialisation. No one has a choice as to the country and circumstances into which they are born. To assign special duties to people on the basis of features of their circumstances over which they have no control at all is deeply unfair.

The APP—ability-to-pay principle—takes an entirely different tack. It does not focus on who has caused the threat to basic rights from climate change, nor on whom is benefitting from it. Instead, it asserts that countries most able to take action to protect the basic rights of people vulnerable to climate change have a duty to do the most. The APP singles out countries that are wealthy, stable and have the capacity to stimulate the creation and rollout of non-carbon based energy sources fit to meet the development needs of present and future people whose basic rights are most vulnerable to climate impacts. A country has the ability to take this kind of action when it can do so without creating new threats to basic rights at home and abroad as a result of that action. On this view, objecting that such action presents a threat to the 'way of life' of a country is irrelevant.

Objections to the APP often turn on how it lets polluters and lucky early industrialisers off the hook in principle. Climate change has been caused by the historical emissions of an identifiable set of countries and people in these countries continue to benefit from these emissions. A failure to make this clear in climate policy now could permit countries such as the USA and members of the EU to unfairly shunt some of the burdens of tackling climate change on to newly industrialising countries of growing wealth such as China and Brazil. The APP fails to correctly identify the source of the special duties of some countries to act on climate change so as to protect basic rights.

We are very far from global action on climate change that reflects any of the principles introduced above. One lesson to distil from these principles for the state we are in is as follows. There is a set of countries that can afford to make radical cuts to the emissions they generate today in order to enjoy the luxuries of consumer society. This set includes countries with high historical emissions, early industrialising countries and rich countries. Members of this set that refuse to sacrifice some of their luxuries, in order that climate-vulnerable people can stay alive, deliberately act in ways that deprive these people of the right to live. In other words, they kill them (Shue, 2001).

Conclusion

The lack of concerted rights-respecting policy on climate change makes it is easy to despair and give up. But there is a different way to see things: could the need for rights-respecting action on climate change be an opportunity for moral progress?

There are two questions we can ask about the vision of an ideal world in which the basic rights of the world's most vulnerable people take centre stage in climate policy (Cohen, 2009). First, how accessible is this state of affairs from where we are now? The scale of the climate challenge and the many obstacles to just climate policy can justify a gloomy answer to the accessibility question. But there is a second question: if we were to achieve this state of affairs where basic rights of the poor took centre stage, how stable would it be? Would the motivations and dispositions of people in a climate-just world sustain the institutions necessary to keep basic rights at the forefront of climate policymaking? Here, there is less cause for gloom. Tackling climate change in ways that protect the basic rights of those least able to protect themselves and least at fault for the threats they face would be a major moral accomplishment for humanity. It would mark out the generation that achieved it as having risen to one of the most dangerous challenges in human history in a way worthy of moral pride. We should articulate the legacy of rights-respecting institutions, policies and political culture that could be left by this action to give a positive answer to the stability question, despite our warranted pessimism about the accessibility of such a world.

NO: Climate change needs a relational, embodied and unbounded perspective
(*Marie-Catherine Petersmann*)

Introduction

> The Anthropocene heralds a rupture within the modern imaginary, calling for modes
> of thinking in obligations beyond the coordinates that have hitherto defined that
> worldview.
>
> (Matthews, 2018)

As alluded to by Matthews in the quote above, answering the question of whether climate change is a human rights violation in the age of **the Anthropocene** requires new modes of thinking about how to live in the present and in the future with new obligations to others. In the first part of this essay, I argue that the international human rights law (HRL) framework is ill-adapted to respond to climate change for three main reasons.

First, a HRL approach to climate change is based on a victim/state binary where the rights holder is the victim and the duty bearer is the state. Fundamental co-responsible duty bearers—such as private corporations and individual consumers—fit uncomfortably in the HRL framework, even though their contribution to climate change is beyond dispute. Second, the individualistic nature of HRL is unsuitable for the collective action problem posed by climate change. The causal link requirement for the admissibility of human rights claims before regional and international human rights courts and tribunals limits the latter's judicial competence to clearly identifiable and directly harmed victims (see **Chapter 13**). The admissibility criteria set under HRL also exclude the extraterritorial implications that states' emissions of GHGs have on victims under the effective control of other states. Finally, the inherently anthropocentric nature of the HRL regime does not account for climate change induced interferences with any non-human and non-animal subjects. The collective action required to combat climate change stands in tension with both the individualistic and anthropocentric normative coding, as well as the admissibility criteria set under HRL.

Drawing on literature from 'new materialism' and Earth system law, the second part of this essay rethinks ways to address climate change from these 'relational' and 'interactional' perspectives. I cannot explore these conceptual models fully here, but instead will focus on their contribution in responding to the shortcomings of the HRL-based approach to climate change argued above by **Catriona McKinnon**.

Not enough: the limits of human rights for tackling climate change

The HRL apparatus posits nation states as *primary duty bearers* and individuals as *right holders*. States have a positive obligation to respect, promote and fulfil human rights. With few exceptions, a sense of 'enchantment' or general optimism is discernible in the literature among those who believe that strengthening HRL can contribute to tackling the problems posed by climate change. Yet the victim/state binary on which HRL relies is unable to account for collective problems such as climate change that affect and are caused by all living beings. This is, to some extent, recognised by human rights experts themselves. The Office of the High Commissioner for Human Rights (OHCHR) notes, for example, that: 'it is virtually impossible to disentangle the complex causal relationships linking historical

[GHG] emissions of a particular country with a specific climate change related effect, let alone with the range of direct and indirect implications for human rights'.[1]

Indeed, climate change results from actions taken by numerous actors across multiple states over extended periods of time. It is therefore impossible to identify one duty bearer and one right holder in a constellation where, to some extent, all states are implicated and where victims are dispersed throughout the globe. Both the causes (GHG emissions) and the effects of climate change (sea-level rise, floods, hurricanes, wildfires, water stress and so on) are transnational. Our conceptualisation of rights and how we seek to operationalise them is therefore not compatible with global problems of such scale (Lofts, 2018: 16–17). What is more, states are not the only responsible agents of climate change and the role of **non-state actors** such as private corporations, investors and consumers in contributing to climate change is beyond doubt (see **Chapter 10** and **Chapter 12**).

The victim/state binary characteristic of HRL means that only victims—or directly harmed individuals who can prove that their rights are infringed—have standing to lodge a complaint against the state in which territory or under whose effective control the violation occurred. Establishing causality in such circumstances requires three steps. First, a state must have contributed to climate change through its actions, omissions or negligence. Second, a specific climate change related harm must have resulted from that state's interference with the climate (see **Chapter 3**). Finally, this specific harm must have directly interfered with the human right(s) of the victim(s). These steps can be a challenging hurdle for claimants to overcome in relation to climate change. Since virtually all states have generated GHG emissions on their territories, the attribution of responsibility gets diluted over time. Given the geographical distances that occur between the source of GHG emissions and their effects, these requirements are close to impossible to fulfil. Thus, as commendable as the argument may be, the HRL paradigm 'cannot address the disjuncture between "victims" and their diffuse or distant "perpetrators" where "violations" are only predicted rather than known and rectifiable' (Tully, 2008: 221).

Likewise, specific climate change related harms that occur today or that will occur in the coming years result from—among other contributing factors—GHGs emitted over the past centuries by a multitude of actors. This reality is unsuited to the temporalities set by HRL. Under HRL, violations are usually established after the fact—once the harm has already occurred—or else high risks and threats thereof are clearly perceptible. In light of its reactive nature, HRL does not account for *prospective* events that can hypothetically harm potential victims. Thus, except at the rhetorical and discursive level, the HRL apparatus cannot effectively respond to projections about future impacts of climate change. While climate change requires anticipated benefits of mitigation—since the measures adopted today to reduce GHG emissions will alleviate climate change in the long run—the *ex post* individualised remedies offered by HRL are unfit for the purpose.

Furthermore, the jurisdiction of HRL instruments is limited to individuals or entities within states' effective control. In other words, states have no justiciable obligations to secure the protection of human rights in relation to climate change beyond their territorial boundaries and jurisdiction (see **Chapter 13**). Historically, developed countries carry the overwhelming share of responsibility for climate change—notably the two hegemonic powers of the nineteenth (Great Britain) and the twentieth (United States) centuries. Together these two countries accounted for 65% of all CO_2 emissions in 1900, 55% in 1950 and still 50% in 1980. Today, China, the USA, India, Russia and Japan together account for nearly 60% of global emissions.

Yet states that are most affected by climate change are often among those whose contribution to GHG emissions is most negligible. Small island developing states (SIDS), for example, face potential inundation by future rises in sea level, while their contribution to GHG emissions is very small. Looking at this issue from a HRL perspective would lead to the absurd scenario where citizens of SIDS can only seek redress for violations of their human rights from their governments, when the latter only contributed negligibly to GHG emissions. These governments are already facing adaptation measures proportionally much greater than states that contributed more to climate change, but are less vulnerable to its immediate effects. Highest emitting states are, thereby, insulated from the human rights claims of individuals living in states that are most vulnerable to climate change. As noted by Tully: 'a human rights orientation will ultimately affirm the primary responsibilities of those States with territorial or jurisdictional control over affected individuals without necessarily enhancing the environmental obligations of other [highest emitting] states' (Tully, 2008: 233).

Finally, individuals living in high emitting states are not all equally empowered to seek redress for violations of their human rights. Climate change affects especially the rights of vulnerable groups—such as indigenous peoples, the world's poor, women and youth—and these adversely affected people(s) might not possess the material and immaterial means necessary to trigger the HRL apparatus. At the same time, in case of a successful HRL complaint, the recognition, reparation and compensation for climate harms would only apply to the victim(s) within the legal jurisdiction. Individual remedy, however, is not sufficient to address the more systemic roots and collective burden of climate change.

Since HRL is not designed as a form of collective power or as a vehicle of popular governance, but only creates individual shields against the exercise of abusive power (Brown, 2004: 461), one must wonder what positive effects—except for the right-holder in question—HRL offers to combat climate change. Parties' argumentations in strategic litigation for climate change increasingly revolve around HRL articulations (Peel and Osofsky, 2018: 37). But the emphasis on the individual human being—or the 'self'—accentuates the rupture between the self and 'others'. These others may be people living in other states or continents, as well as other living and non-living beings not endowed with 'rights'.

Indeed, as HRL is by its very nature strictly anthropocentric—centred around the human being—non-human and non-animal beings fall outside the scope of consideration whenever they do not directly impact the human right(s) of the victim(s). If the quality of the human environment is taken into account by human rights courts when it directly interferes with the right(s) of the applicant(s), no HRL instrument provides an unconditional and justiciable protection of nature for its intrinsic value. In doing so, the HRL framework perpetuates the erroneous idea that the environment refers to something that surrounds and serves the human at its centre (Petersmann, 2018). If a stable climate cannot be effectively protected under a HRL framework, what alternative conceptual models exist to tackle climate change?

Beyond human rights 'enchantment': new materialism and Earth system law

While HRL has seemingly little to offer in effectively combatting the climate catastrophe we face, HRL language and institutions have monopolised global imagination among those committed to action. Since the mid-2000s, the HRL rhetoric has become prominent in relation to climate change. A form of psychological 'enchantment' prevents international lawyers—and human rights lawyers in particular—from looking at global problems through any other prism than international (human rights) law. For Koskenniemi, 'lawyers are enchanted by the law that is familiar to them and the institutions and practices they are involved with; that makes them often unable to find a good solution to the problem they are faced with' (Koskenniemi, 2018).

This enchantment towards international HRL in international environmental law (IEL) —or the mainstreaming of HRL (Koskenniemi, 2010) in IEL—blinds international human rights and environmental lawyers to more suitable ways of re-envisioning our responses to climate change. If we take the critique set out above seriously, climate change cannot be addressed by only thinking about the 'self'—or one's own living conditions and interests framed as 'rights'—in light of the actions or omissions of one's state of residence and/or nationality. Instead, what is needed is a reconceptualisation of the currently predominant public law and state-centric approach to climate change that underpins both HRL and IEL. In this approach, the concepts of sovereignty, strict causality and *ex post* temporality are limiting meaningful, adaptive and foreseeable action.

The multidisciplinary field of *environmental humanities* provides important insights into how to respond to the shortcomings of a HRL-based approach to climate change. Environmental humanities address ecological problems from closely knit ethical, cultural, philosophical, legal, political, social and biological perspectives (Iovino and Oppermann, 2016: 1). Environmental humanities bridge the traditional dualisms between the social and the natural sciences, as well as between western and non-western and indigenous ways of relating to the natural world. Various theories have been developed that transcend the geographic and jurisdictional boundaries essential to HRL. These also transcend the human-nature divide—where humans are seen as subjects and the environment as a surrounding object—thereby decentring humans' rights in tackling climate change.

Following an ontological turn to materiality[2] in anthropology and philosophy (Vermeylen, 2017: 141), **new materialists** are calling for a similar emphasis on material objects across the humanities and social sciences, including in international (environmental and human rights) law. This has brought about a renewed interest in the embodiment of humans in a material world. The material world has for a long time been the central focus of **Actor-Network Theory** (Latour, 2005), which describes how the natural and the social worlds exist in constantly shifting networks of relationships. Through this prism, the dichotomy between nature and society—between human and non-human—that structures liberal legal imaginaries is suspended and traded for a relational ontology. New materialists emphasise humans' embeddedness in more-than-human natural processes and the inter-connectedness between all 'actants' (Latour, 2005: 71) as part of heterogenous assemblages. From this perspective, human laws—including HRL and IEL—are viewed as emerging from and situated in 'socionatures' or 'natureculture'. These notions posit the inseparability of nature and culture in ecological relationships that are both biophysically and socially formed (Haraway, 2003).

This approach strongly resonates with vulnerability theory and 'relational law'. An ecological vulnerability frame emphasises the 'vulnerability of the entire living order' (Kotzé, 2019: 62) and the 'interdependency of the human body with a complex array of nonhuman and trans-human systems' (Harris, 2014: 126). These frames require us to think about the embodied nature of the individual—or the individual's material embodiment in socionature. They conceive of an individual's protection from a relational perspective that takes account of not only the individual, but also its web of vital relationships.

As the definition of the word indicates, to 'relate' does not only mean 'to stand in relation to' but also 'to feel sympathy for or identify with'. A relational approach accounts for a much wider set of relations in comparison to the strict causality demanded by HRL. Similarly, since all states share an ecological vulnerability—even though they are differently vulnerable to climate harms—a relational law perspective revisits the notion of state sovereignty that is central to a HRL-based approach to climate change. HRL invokes and reinforces the image of the territorially bounded autonomous liberal state by describing the scope of state obligations as (extra)territorial. In contrast, relational law views states' obligations as shared in light of responsible relationships across borders, whether transboundary, transnational or global. It highlights the necessity of international cooperation for problem solving (Seck, 2019: 176).

Based on similar premises, *Earth system law* is another conceptual model that comprehends the Earth system as consisting of complex, evolutive and adaptive biophysical and socio-political arrangements that are inextricably interconnected (Kotzé, 2019). Here, the human-environmental relationship is underpinned by an 'interactional' ontology, in which social and ecological components are considered to be distinct, but interacting within a system. From an Earth system perspective, climate change should not be tackled in isolation, but through integrated modes of polycentric, reflexive and multi-scalar global law and governance (Biermann, 2014). Earth system law therefore embraces complexity, instability and unpredictability of socio-ecological systems, against the traditional legal precepts of order, certainty and predictability. Some argue in favour of enhancing the resilience of socio-ecological systems (Folke, 2006: 259). This entails an enhancement of the capacities of socio-ecological systems or assemblages to transform themselves in light of external stresses and to adapt to constant non-linear and unpredictable changes. What matters, in essence, is not the security and stability of individual parts within a system—as is the case from a HRL perspective. Rather what matters is the system's capacity to learn from previous experience and to take pre-emptive measures to address prospective changes.

These relational and interactional paradigms provide new doctrinal and technical tools to address climate change. They overcome the false dichotomies between society and nature, subject and object, animate and inanimate and duty bearer and rights holder, that are entangled in HRL discourse and practice. Here, all living and non-living subjects—as interconnected and co-responsible actants within the Earth system—play a role, regardless of a direct causal relationship to the harm caused or endured. From this perspective, we move past the rigid and narrow victim/state binary of HRL into a 'dis-anthropocentric' alliance of entangled agents that work with, through and across material agencies that comprise the world (Iovino and Oppermann, 2016: 13).

Conclusion

The 2015 Paris Agreement refers to climate change not as a human rights violation, but as a 'common concern of humankind'. This does not mean that climate change has no impact on human rights. That climate change will lead to many human rights infringements is no longer debated. Against this background, recalling the negative impacts that climate change has on human rights is key to articulating a moral or normative case for action on climate change. What is debated, however, is whether climate change should be *framed* as a human rights violation.

In this essay, I have argued that addressing climate change through a HRL prism is counterproductive in light of HRL's individualistic nature, its strict admissibility criteria and its inherently anthropocentric frame. HRL oversimplifies the causally complex problems posed by climate change by being too reductive in its reliance on a vocabulary of 'violators' and 'victims' (Pedersen, 2010: 250). The universalisation of HRL as a framework to pressure states to combat climate change has diverted resources and focus away from radically different and more productive strategies capable to bring about Earth system justice in relation to climate change. Against this backdrop, more voice and space should be granted to discourses stemming from environmental humanities and new materialist theories. These resist the modern binaries of human/non-human, nature/society, subject/object, animate/inanimate or state/victim which circumscribe the western legal imagination.

Imperative calls for action against climate change need not necessarily be framed as rights. Other paradigms that acknowledge and emphasise the relational, embodied and unbounded situation we are in can offer more appropriate and 'down to earth' vocabularies and technologies to orient ourselves in this new climatic regime (Latour, 2018). To promote Earth system justice, a change in how we use the law—and for what purpose— is required. We need one that abandons the sovereign-centric and territorially and human-bounded approach characteristic of HRL and shifts towards legal materialism, vulnerability theory and relational law. Instead of hubristic faith in techno-managerial salvation—in light of which climate change could be solved through science and technology—more humility is required to apprehend our life as part of nature (Hulme, 2014). In Argyrou's powerful words:

> [w]hat is needed above science is something that captivates people's full being—a system of values, a moral story, an ontological master narrative within which the ecological crisis becomes not only visible but also relevant and meaningful.
>
> (Argyrou, 2005: 48)

As argued in this essay, this further necessitates the acknowledgement of both our vulnerability to and our dependency upon a complex web of material relations. Humbly embracing these two premises leads us to revisit central paradigms of order, certainty and predictability of and within the law. The material networks we are part of are in constant evolution and so should our understanding of law and justice. They should strive for frequent adjustment to the perpetual transformation of the human and non-human networks we are embedded in that determine Earth's life support systems.

Further reading

Bell, D. (2013) Climate change and human rights. *WIREs Climate Change.* **4**(3): 159–170.

This review article explains the attractions of a human rights approach to climate change. Bell proposes three main arguments connecting human rights and climate change: that there is a human right to a stable climate; that anthropogenic climate change violates basic human rights to life, health and subsistence; and that there is a human right to emit greenhouse gases.

Climate Equity Reference Project. https://climateequityreference.org/

This website explains this long-term initiative which provides scholarship, tools and analysis to advance global climate equity—as a value in itself and as a realistic path towards an ambitious global climate regime. The website also includes an on-line Climate Equity Reference Calculator that allows users to interactively explore user-defined implementations of the global climate effort-sharing framework.

Dietzel, A. (2019) *Global Justice and Climate Governance: Bridging Theory and Practice*. Edinburgh: Edinburgh University Press.

This book evaluates the global response to climate change from a cosmopolitan justice perspective. Investigating the role of states, cities, corporations and NGOs in the post-Paris Agreement era, Dietzel illustrates that climate justice theory can be used to assess and compare both state (multilateral) and non-state (transnational) climate change governance—in other words, that theory and practice can be bridged.

Humphreys, S. ed. (2009) *Human Rights and Climate Change*. Cambridge: Cambridge University Press.

This inquiry into the human rights dimensions of climate change looks beyond the potential impacts of climate change to examine the questions raised by climate change policies: accountability for extraterritorial harms; constructing reliable enforcement mechanisms; assessing redistributional outcomes; and allocating burdens, benefits, rights and duties among perpetrators and victims, both public and private.

McKinnon, C. (2012). *Climate Change and Future Justice: Precaution, Compensation, and Triage*. London: Routledge.

Climate change creates unprecedented problems of intergenerational justice. This book explores the question: What do members of the current generation owe to future generations in virtue of the contribution they are making to climate change? McKinnon applies the important principles of democratic equality to the most serious set of political challenges ever faced by human society.

Wewerinke-Singh, M. (2019) *State Responsibility, Climate Change and Human Rights under International Law*. Oxford: Hart Publishing.

The Paris Agreement is the first multilateral climate agreement to refer explicitly to states' human rights obligations in connection with climate change. This book offers an analysis of the legal issues related to accountability for the human rights impact of climate change. It explains when and where state action relating to climate change may amount to a violation of human rights and evaluates various avenues of legal redress available to victims.

Follow-up questions for use in student classes

1. What protection does the 2015 Paris Agreement on Climate Change give to people's basic rights in the face of worsening climate change?
2. Why do we think that we are required to respect the basic rights of future people when these people do not yet exist?
3. Why are the admissibility criteria set under the human rights law framework difficult to fulfil for alleged climate change-induced human rights violations?
4. Does human rights law provide effective remedies to climate change impacts—such as sea-level rise in small island developing states—caused by greenhouse gases emitted by China or the USA?

Notes

1 Report of the OHCHR on *The Relationship Between Climate Change and Human Rights*, UN Doc.A/HRC/10/61 (15 January 2009), para 70.
2 The 'ontological turn to materiality' refers to a theoretical movement in the social sciences which emphasises the study of material objects, instruments and embodiments in how ideas and institutions come into being and influence.

References

Argyrou, V. (2005) *The Logic of Environmentalism: Anthropology, Ecology and Postcoloniality.* New York/Oxford: Berghahn Books.

Armstrong, C. (2019) *Why Global Justice Matters: Moral Progress in a Divided World.* Cambridge: Polity Press.

Biermann, F. (2014) *Earth System Governance: World Politics in the Anthropocene.* Cambridge, MA: MIT Press.

Brown, W. (2004) 'The most we can hope for … ': human rights and the politics of fatalism. *South Atlantic Quarterly.* **103**(2–3): 451–463.

Campbell, T. (2019) Human rights. In C. McKinnon, R. Jubb and P. Tomlin, eds. *Issues in Political Theory.* Oxford: Oxford University Press. pp.220–240.

Cohen, G.A. (2009) *Why Not Socialism?* Princeton, NJ: Princeton University Press.

Folke, C. (2006) Resilience: the emergence of a perspective for social-ecological systems analyses. *Global Environmental Change.* **16**(3): 253–267.

Gardiner, S.M. (2011) *A Perfect Moral Storm: The Ethical Tragedy of Climate Change.* Oxford: Oxford University Press.

Haraway, D.J. (2003) *The Companion Species Manifesto: Dogs, People, and Significant Otherness.* Chicago, IL: The University of Chicago Press.

Harris, A. (2014) Vulnerability and power in the age of the Anthropocene. *Washington and Lee Journal on Energy, Climate and Environment.* **6**(1): 98–161.

Hulme, M. (2014) *Can Science Fix Climate Change? A Case against Climate Engineering.* Cambridge: Polity Press.

Iovino, S. and Oppermann, S. (2016) Introduction. In S. Iovino and S. Oppermann. eds. *Environmental Humanities: Voices from the Anthropocene.* Lanham, MD: Rowman and Littlefield. pp.1–21.

Koskenniemi, M. (2010) Human rights mainstreaming as a strategy for institutional power. *Humanity, an International Journal of Human Rights, Humanitarianism and Development.* **1**: 47–58.

Koskenniemi, M. (2018) Sovereignty, property, and the locus of power [online]. *JHIBLOG: the blog of the Journal of the History of Ideas.* Accessed 2 June 2019. Available from: https://jhiblog.org/2018/10/17/sovereignty-property-and-the-locus-of-power-anne-schult-interviews-martti-koskenniemi-on-the-conceptual-history-of-international-law/

Kotzé, L.J. (2019) The Anthropocene, earth system vulnerability and socio-ecological injustice in an age of human rights. *Journal of Human Rights and the Environment.* **10**(1): 62–85.

Latour, B. (2005) *Reassembling the Social: An Introduction to Actor-Network-Theory.* Oxford: Oxford University Press.

Latour, B. (2018) *Down to Earth: Politics in the New Climatic Regime* (trans. C. Porter). Cambridge: Polity Press.

Lofts, K. (2018) Analysing rights discourses in the international climate regime. In S. Duyck, S. Jodoin and A. Johl, eds. *Routledge Handbook of Human Rights and Climate Governance.* Abingdon: Routledge. pp.16–30.

Matthews, D. (2018) Obligations in the new climatic regime [online]. *Critical Legal Thinking.* Accessed 2 June 2019. Available from: http://criticallegalthinking.com/2018/07/16/obligations-in-the-new-climatic-regime/

Pedersen, O.W. (2010) Climate change and human rights: amicable or arrested development? *Journal of Human Rights and the Environment*. 1(2): 236–257.

Peel, J. and Osofsky, H.M. (2018) A rights turn in climate change litigation? *Transnational Environmental Law*. 7(1): 37–67.

Petersmann, M.C. (2018) Narcissus' reflection in the lake: untold narratives in environmental law beyond the Anthropocentric frame. *Journal of Environmental Law*. 30(2): 235–259.

Seck, S. L. (2019) Relational law and the reimagining of tools for environmental and climate justice. *Canadian Journal of Women and the Law*. 31(1): 151–177.

Shue, H. (1996) *Basic Rights: Subsistence, Affluence, and U.S. Foreign Policy*. Princeton, NJ: Princeton University Press.

Shue, H. (2001) Climate. In D. Jamieson, ed. *A Companion to Environmental Philosophy*. Oxford: Blackwell. pp.449–460.

Tully, S. (2008) Like oil and water: a sceptical appraisal of climate change and human rights. *Australian International Law Journal*. 15: 213–233.

UNDP, ed. (1994) *Human Development Report 1994*. New York: Oxford University Press.

Vermeylen, S. (2017) Materiality and the ontological turn in the Anthropocene: establishing a dialogue between law, anthropology and eco-philosophy. In L. J. Kotzé, ed. *Environmental Law and Governance for the Anthropocene*. Oxford: Hart Publishing. pp.137–162.

Part IV

Who should be the agents of change?

12 Does successful emissions reduction lie in the hands of non-state rather than state actors?

Liliana B. Andonova and Kim Coetzee

Summary of the debate

This debate concerns the relative efficacy of state and non-state actors in delivering the scale and speed of greenhouse gas emissions reductions deemed necessary to stabilise the climate. **Liliana Andonova** argues that increasing numbers of non-state actors—companies, associations, advocacy organisations, cities, regions—are, *de facto*, taking on formal and informal responsibilities for reducing emissions of greenhouse gases. The implementation of emissions reductions implies a behavioural change across industry and society and so non-state action is essential for effective emissions reduction. Alternatively, **Kim Coetzee** argues that states still hold sway over significant regulatory, investment and behavioural policies that hold more promise for enabling far-reaching change. The changes to modern living which are required are of such a nature and scale that state actors are best positioned to intervene directly at a system-level.

YES: Because it requires commitments by all actors, private and public
(*Liliana B. Andonova*)

Introduction

Direct efforts by non-state actors to reduce their carbon footprint are essential for climate mitigation. Climate change is a problem of formidable complexity. It results from multiple streams of socioeconomic activities and their interface with complex Earth system processes. The impacts of climate change vary across regions and localities. The response, therefore, requires mobilisation of actors across all sectors of society and levels of governance in support of a series of targeted interventions and innovations. A single intergovernmental treaty and the willpower of states to implement it are alone unlikely to provide sufficient solutions for an issue of such complexity.

This essay advances the argument that if we conceive of climate change in planetary terms, as a problem with many layers that vary across geographies and scales of social organisation, successful emissions reductions lie to a large degree in the hands of non-state actors. **Non-state actors** are defined broadly as entities whose action on climate change is not directly mediated by the central government. They may include businesses, civil society organisations, cities, regions, universities, foundations, associations and so forth. During decades of governmental gridlocks, many non-state actors have already shown willingness to engage in climate initiatives within their core activities or through transnational networks (Hoffmann, 2011). Each non-state initiative may have a small or even negligible immediate impact in solving the planetary challenge; yet, each initiative also builds capacity and political constituencies to demand stronger commitments and governmental regulation.

Non-state initiatives break up the planetary climate change problem into smaller components, in search of experimental solutions with better fit to scale and context. They open possibilities for learning-by-doing and catalytic effects. Finally, industries, markets and social entrepreneurship are likely to be necessary mechanisms in the engine of innovation for climate mitigation and the diffusion of new technologies. Therefore, the direct engagement of non-state actors is essential and urgent. There is no time left to keep waiting for the trickle-down effects of a global treaty and its related government regulations.

Framing climate change

Since climate change gained political saliency in the late 1980s, it has been predominantly viewed as a **tragedy of the commons** dilemma on a planetary scale (see **Chapter 14**). From this perspective, states are the main actors. The solutions that have been attempted are intergovernmental treaties, which have established a framework of broad common but differentiated responsibilities (the UN Convention on Climate Change (UNFCCC), 1992), partial emissions-reduction targets (the **Kyoto Protocol**, 1997) and subsequently a treaty embracing **nationally determined contributions** (NDCs) for emissions reductions for all signatories (the **Paris Agreement**, 2015). Yet, more than 30 years after the problem of climate change was placed on the international agenda, we witness still very limited state ambition or implementation of emissions reductions. Climate mitigation, a global public good that is widely diffused across jurisdictions and generations, requires long-term reciprocity by all significant emitters of greenhouse gases (GHGs). However, ambitious government strategies and reciprocal commitments—including by major economies such as the United States, India, China, Brazil, Russia, Australia and Canada—have been pre-empted by political concerns about economic competitiveness, development priorities, domestic interests, normative dissonances and distributive justice (see **Chapter 1**).

Fortunately, the idea of climate change as a '**wicked problem**' that defies collective action is challenged by alternative framings of climate change as a complex problem of interacting socio-economic systems and Earth system processes. In other words, because it is characterised by multiple sources and manifestations that vary across sectors, geographies, ecological systems, economies and actors, climate change is not a *single* problem (Keohane and Victor, 2011; Ostrom, 2014). At the same time, these variations also influence the benefits of mitigation and the beliefs about the synergies and trade-offs of low carbon development. A perspective which portrays climate change as a complex and distributed issue implies the necessary intervention of all relevant actors for effective emissions reductions to be delivered, in a polycentric mode of governance (Ostrom, 2014).

Non-state action when states lag behind

Non-state actor initiatives on climate change came into prominence during the long decade of intergovernmental gridlock that began in the late 1990s shortly after the adoption of the Kyoto Protocol (Bulkeley et al., 2014). Transnational governance networks engage non-state actors such as companies, industry associations, advocacy organisations or expert organisations, as well as sub-state actors such as regions, cities or public administration departments across borders to advance informal rules and behaviour change towards shared objectives of public significance (Andonova et al., 2017; Bulkeley et al., 2014). Transnational climate governance thus does not directly depend on the diplomatic intermediation of the central state, but rather involves entrepreneurial initiative from a multitude of actors (Andonova, 2017; Green, 2014). Figure 12.1 captures the trend of some 119 transnational climate initiatives documented in the Transnational Climate Database (see Bulkeley et al., 2014, with updates by Andonova et al., 2017; Michaelowa and Michaelowa, 2017).

These data reveal a great diversity of non-state action with variable structure and focus. A large number of initiatives are largely private, in the sense that they are organised by, and seek to steer the actions of, non-state actors. For example, the Global Sustainable Electricity Partnership (GSEP) is a consortium of leading electricity companies. It advances commitments by the top management of its members to increase the share of renewable energy sources in electricity production and consumption (see **Chapter 7**), including through project development and technology diffusion in developing countries. The Carbon Disclosure Project (CDP) has provided a platform for companies, and subsequently for cities, to disclose their carbon emissions and incentivise emissions monitoring and reduction through systematic information and benchmarking.

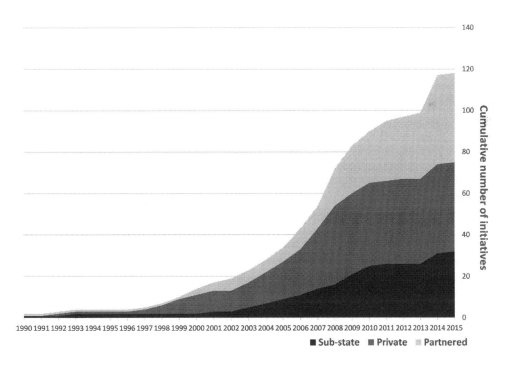

Figure 12.1 Engagement of non-state actors in transnational climate governance

Sources of data (Andonova et al., 2017; Michaelowa and Michaelowa, 2017).

A growing number of transnational initiatives for climate change take place through partnerships between non-state actors and more traditional intergovernmental organisations or public agencies. Public-private collaboration varies in focus, including the promotion of investment and access to cleaner energy technologies, support for the development of carbon markets (see **Chapter 6**), or the creation of funding instruments for project-based offsets. Cities and other sub-state actors have increasingly engaged in networks across borders, such as the Local Governments for Sustainability network (ICLEI), **Regions 20** or the **Covenant of Mayors**, to leverage resources for climate change mitigation and adaptation through improved infrastructure, energy efficiency and urban planning. Advocacy organisations have pursued multiple strategies, including direct lobbying, but also the creation of standards for the regulation of voluntary carbon offsets in order to incentivise greener projects and the promotion of sustainable development co-benefits and adaptive capacity.

While Figure 12.1 captures the tendency for proliferation and diversity of non-state initiatives for climate mitigation and adaptation, the literature has illuminated a variety of sector-specific and place-specific incentives and ideologies that drive non-state engagement. These range from gaining a first-mover advantage in the development of new carbon and technology markets, to fostering linkages between climate change and local priorities such as advancing resilience, mobilising project-based investments or leveraging resources for sustainable urban management (Andonova et al., 2017; Bulkeley et al., 2014; Green, 2014; Hoffmann, 2011). Would such diverse, decentralised and dispersed actions generate sufficient emissions reductions to have a tangible impact on climate stabilisation? The verdict is still out. Michaelowa and Michaelowa (2017) argue that in the short-term this is unlikely. At the same time, a study by the UN Environment Programme tested a methodology to quantify the impact of 15 transnational actor initiatives. The study estimated that these actors would reduce by approximately three billion tonnes GHGs emitted into the atmosphere by 2020—more than the annual emissions of India and not dissimilar in magnitude to the four to six billion tonnes the UN estimates that national pledges announced in Paris will remove by 2030.[1]

Ultimately, the most immediate and important effects of the upsurge of such voluntary non-state initiatives are political and social. They create new constituencies across sectors of society and in public administrations that are able to ratchet up demand for more effective institutional solutions (Bernstein and Hoffmann, 2018; Hale, 2016). They also undermine the false ideational dichotomies between development and decarbonisation, between economic progress and climate mitigation and between local benefits and global concerns. This is accomplished through the practices and advocacy, working across national borders, of non-state actors in industry, agriculture, forestry, finance, health, telecommunications and urban planning. The Paris Agreement has formally recognised the policy significance of non-state actors in the all-out effort on climate change (Hale, 2016). Other institutional developments, such as the adoption of Sustainable Development Goal 7 on the advancement of and access to clean energy, or the creation of national carbon-trading platforms in large emerging economies such as China (see **Chapter 14**), would not have been politically feasible without almost two decades of transnational project-based offsets, clean energy investments and capacity-building activities spearheaded by transnational non-state governance.

Climate experiments and catalytic effects

The agency behind non-state action for climate mitigation differs considerably from that of states, which are primarily responsible for formal international agreements, national policies and their implementation, as argued below by **Kim Coetzee**. Frequently, if not always, it is governance entrepreneurs—actors actively seeking and prepared to take risks to spur ideational, societal or institutional change—that initiate voluntary climate action (Andonova, 2017; Green, 2014). Such entrepreneurial schemes are typically experimental in their initial stages. They allow for testing new ideas and market-based or institutional instruments at different entry points of the climate dilemma, something which could open pathways towards place-adapted and sector-specific decarbonisation (Bernstein and Hoffmann, 2018).

In 2003, for example, the Mayor of the Galápagos Island of San Cristóbal engaged in a partnership with a number of global electricity companies (under the GSEP umbrella) to install wind power technology that would displace 50% of the island's dependence on diesel fuel. The municipality sought to address urgent local problems such as import dependence, air pollution and the risks of tanker accidents and fuel leaks that contaminate the pristine ecosystem and threaten marine wildlife. For the industry, the project presented both risks and the opportunity to build the first-ever wind park located in a remote island environment and not connected to a larger grid. This type of experimentation, with the deployment of new technologies and the associated level of infrastructure development, domestic capacity and social acceptance, thus became possible because of the willingness of local and industrial actors to undertake a risk. It was also enabled thanks to the political and fiscal support lent to the partnership by the national government, the UN Development Programme and, in view of its global significance, the UN Foundation.

Bernstein and Hoffmann (2018) and scholars of innovation (e.g. Anadon et al., 2016) have argued that such experiments are essential in systems characterised by high complexity, since linear solutions are unlikely to be sufficient for adaptive learning and disruptive breakthroughs. Private and transnational initiatives are less likely than state governments to propose regulatory solutions (Bulkeley et al., 2014). They tend to diffuse information, technology and capacity for innovation, learning-by-doing and adaptation at multiple levels. Ultimately, however, experiments need to have some type of a catalytic or ripple effect to produce an impact on a planetary system. The agency of both private and public actors in identifying promising experiments and investing political, social and environmental resources for their scaling-up, remains critical for emissions reduction and climate stabilisation (Hale, 2016).

Innovation and markets

The reversal of GHG emissions trajectories to prevent catastrophic climate change requires significant innovation in technologies and processes, as well as their adoption in societies with different development, institutional and resource characteristics. It would be naïve to believe that such a transformation could be achieved through the magic wand of intergovernmental treaties, government regulation or carbon taxes alone, even if regulatory politics were to improve and make such a package politically plausible. The International Renewable Energy Agency reports that the significant growth in the deployment of renewable energy technologies in the last decade has been financed primarily by private investment. In 2016, private sources provided 90% of the investment in renewable energy.[2] Bettencourt et al. (2013) document a sharp increase in renewable technology innovation globally between 2004 and 2009, despite low and stagnating levels of public R&D funding. Their model attributes this spur in innovation to growing private markets as 'vital complements' to public R&D.

Public policies such as those adopted by the European Union, and more recently in large emerging markets such as China and India, have provided an important signal and incentives for the development and deployment of renewable energy. Yet it is global markets, foreign direct investment (FDI) and expanding production capacity that have been among the main drivers of the decrease in the price of renewable energy and the subsequent expansion in global access to these technologies (see **Chapter 7**). Even the humble LED lightbulb has diffused largely through a combination of policies, FDI, growing production capacity and trade. While state-driven policies provide important signals for clean technology innovations, markets and private incentives of industry and consumers are powerful drivers of the diffusion of low carbon technologies (see **Chapter 6**). Indeed, facilitated access to cleaner technologies and investment can provide an additional rationale for governments in developing countries to adopt more progressive climate policies.

Societal actors, human capital and institutions are in turn essential for the domestication of new technologies and the creation of innovation systems that are adapted to local scales and conditions. The documented difficulties of international efforts to diffuse solar and other low-emission cook stoves in rural developing areas, despite obvious damage to health and the environment from existing open-fire technologies, have been a wake-up call about the complexity and social embeddedness of innovation processes. As analysts have recently argued in their reflection on cross-sectoral innovations for sustainability, 'technological innovation processes … emerge from complex adaptive systems involving many actors and institutions operating simultaneously from local to global scales' (Anadon et al., 2016: 9682). Non-state actors from entrepreneurs to technology users, universities, companies, institutional investors and civil society organisations are at the core of dynamic innovation systems. These have the potential to disrupt the lock-in of carbon intensive technology and to scale-up emissions reduction solutions.

Conclusion

This essay conjectures that the successful reduction of GHG emissions depends to a great extent on the incentives, entrepreneurial initiatives, investments and direct mitigation efforts of non-state actors. But is climate stabilisation entirely in the hands of private actors *rather* than the state?

Emphatically, no! The argument developed here does not imply a zero-sum trade-off or a functional equivalence between non-state action and public regulations. On the contrary, it calls for understanding the climate change dilemma in its full planetary and social complexity. Such an understanding leads us to challenge false dichotomies between public and private actors, between markets and regulations, between Earth systems and the economy and between intergovernmental agreements and decentralised local, regional or private solutions. Nobel Prize winner Elinor Ostrom was one of the first scholars to call our attention to the polycentric nature of the climate problem and the necessity of a polycentric system of governance. Ostrom (2014) also warned that there are 'no panaceas', no magic bullets that will provide a unique institutional solution to a system of such complexity. A set of loosely coupled public, non-state and hybrid institution and commitments are likely to be necessary. As Hale (2016) observes, the Paris Agreement requires countries and societies to put 'all hands on deck' in reducing emissions and stabilising the climate system.

The argument developed here therefore posits that non-state action is essential for emissions reduction and also interdependent with that of the state, rather than a second-best or contradictory pathway. This implies that non-state actors can no longer hide behind the low ambition of present intergovernmental agreements. Non-state actors themselves have the responsibility and agency to demand strong public policies and to directly invest material and non-material resources in economic and social change.

Finally, we must raise the question of accountability which is presently the missing link for both private and public climate commitments. Civil society organisations, municipal offices, carbon trading associations, audit and accounting firms, insurance companies and public agencies can and should be part of a robust system of climate accountability, which requires societal fire-alarm mechanisms and public scrutiny. Accountability reduces the risk of greenwashing and empty promises by private and public actors. It increases societal trust and the likelihood of diffusion of climate experiments and innovation and their subsequent catalytic effects. Reducing emissions requires an all-out effort and societal actors are at the centre of it.

Acknowledgements

I am grateful to Mike Hulme for the generous guidance and feedback and to Dario Piselli for invaluable research and comments. The support of the Swiss Network for International Studies (SNIS project 3369) is gratefully acknowledged.

NO: Successful emissions reduction requires a system-level response that states are best placed to facilitate (*Kim Coetzee*)

Introduction

The observation that state actors have not done enough to implement responses to climate change since multilateral negotiations began in the 1990s does not absolve states from the necessity or responsibility to do so. The world's historical and continued reliance on fossil fuels to create the energy that powers modern economies has already caused global warming of approximately 1°C since pre-industrial times. According to the IPCC's Special Report on 1.5°C (SR15), at the present rate of GHG emissions Earth is expected to have warmed to 1.5°C above the pre-industrial temperature by 2040 (IPCC, 2018). SR15 also indicated that emissions would need to decline precipitously by 45% between 2010 and 2030 and must ultimately reach 'net zero' by 2050 in order to keep the global temperature increase to 1.5°C. Given that energy is the backbone of modern societies, and the SR15 timeframes are shorter than those needed to keep below the Paris Agreement's goal of a 2°C increase, only wide-ranging, systemic changes to national economies will bring about the rapid decline in emissions necessary.

This essay argues that, in the current international system, state actors are still best positioned to undertake or facilitate the radical, institutional and system-level changes required at the speed and scale necessary to keep warming to within 1.5°C. Non-state actors undoubtedly have a role to play in building the response to climate change, as argued above by **Liliana Andonova**, but it is not a sufficient one. The following paragraphs explain the ways in which state actors remain crucial to shaping the national policy landscapes within which individuals and other, larger, non-state actors operate.

States, and only states, can implement the policies, regulations and incentives that alter calculations of costs and benefits which in turn constrain and enable non-state actors' actions with respect to GHG emissions.

States: down, but not out

Globalisation and the growing importance of subnational and transnational actors have challenged the dominance of the state in world affairs. But despite the evolution of multilevel governance practices, the international system remains a state-centric one. In the current international system states, as legal entities, retain the sole authority to enter into treaties with other states. In the absence of an overarching authority at the international level which could compel states to act—the fundamental *problematique* of international relations—treaties between states remain the primary formal mechanism through which international stability is achieved and maintained. These treaties and agreements then in effect become an overarching framework within which all signatory states operate. Treaties are therefore an essential component of the architecture of international relations. With respect to climate change, that framework began with the UNFCCC in 1992 and was followed by the Kyoto Protocol in 2005 and the Paris Agreement in 2015.

There is some disagreement between authoritative voices in the international climate change policy field as to whether the Paris Agreement as a whole is a formal, binding legal treaty or simply an international agreement (see **Chapter 13**).[3] Contestation aside, however, the Paris Agreement contains both legally binding and non-binding provisions. States have agreed to bind themselves to measuring and reporting on their activities (transparency provisions), but there are no legally binding provisions relating to emissions reduction goals. These goals are voluntary and known as **Nationally Determined Contributions** (NDCs). States have undertaken to revisit their 2015 pledges by 2020 and put forward more ambitious NDCs every five years thereafter. Theoretically, this will be a mechanism to increase emissions reduction ambition. By signing up to the Paris Agreement, states also agreed to re-assess the cumulative reduction in emissions every five years (accountability or stock-taking provisions). Finally, as the parties to the negotiations could not come to an agreement on censuring under- or non-performing states, there are no formal penalties or consequences for states that fall short of their own NDC goals.

Safety in numbers

The transparency and stocktaking provisions of the Paris Agreement do, however, provide a measure of certainty for all states involved—as international agreements and institutions are designed to do. Being reasonably confident that all states are taking *some* measures to reduce emissions in the form of NDCs means that all states can potentially implement measures which might be costly, but which will ultimately reduce emissions. The Paris Agreement therefore provides some assurance that no state bears all the risks and costs inherent in acting unilaterally or first. In this way, and in the absence of an overarching authority, the Agreement has helped to change the assessment of the costs and benefits of policy action for each state.

In contrast, at the national level states *are* the overarching authority within their jurisdiction and therefore have the authority to create legal frameworks that bind their citizens and other non-state actors to act. Legal frameworks may include laws, regulations and incentives, as well as long-term policy and development goals (see **Chapter 14**). Hence, with the 'safety' of some mutual understanding in place at international level, states can play a facilitative role within their territorial borders by creating a framework that alters the assessment of the costs and benefits of working to reduce emissions for citizens and other non-state actors.

Is responding to climate change at an individual level even rational?

Virtually every aspect of modern living is energy dependent. From the food we eat, the water we drink, to the clothes we wear, the packaging we discard, the housing that shelters us, the bitcoin we buy, the transport we use to do the work we do... energy is required to make, distribute and even eliminate all or part of these goods. For most states this energy is still predominately derived from fossil fuels. Reducing one's individual carbon footprint in light of the institutionalised embeddedness of fossil fuels within modern economies can seem overwhelming. In the absence of an external force that changes that calculation of costs and benefits there is limited incentive for people to change their everyday decisions at the individual level.

For example, for an individual the most logical and economically rational response to a global commons problem like climate change is to do nothing to reduce the impact their lifestyle choices (what they eat, how they travel and so on) have on others or on the environment. Each individual is effectively enjoying the benefits of living in a fossil-fuelled economy, without (usually) paying for the social or environmental damage caused to others (both currently alive and future humans) and the environment by their choices. The costs of their actions have been externalised (see **Chapter 5**). Any one individual's efforts to reduce their personal emissions ultimately has a very limited impact on overall cumulative global emissions (Sinnott-Armstrong, 2005).

Let me give a personal example. Should I choose to disengage from the fossil fuelled economy for reasons of my own by, for example, walking everywhere instead of driving or catching a bus, I alone will 'pay' for this decision in terms of the extended time I spend commuting, wearing through my shoes more quickly, inhaling traffic fumes, and so on. The people driving past me in *their* vehicles however, are not negatively affected by my choice to opt out of using petrol, diesel or gas transport in order to reduce my carbon footprint. At least, certainly not in the way that *I* am directly affected by *their* vehicles' emissions as they drive past me or that I will eventually be affected by *their* contribution to causing a changing climate. Framed this way, it makes the most rational sense for me to continue using fossil fuels to maintain my lifestyle and or achieve my aspirations unless, or until, *all* other actors in the economy—or at least a very significant number of them—are compelled to stop using fossil fuels too.

It is at this point that states play a role that most non-state actors cannot—theoretically they have the authority to compel all their citizens to comply with initiatives to reduce emissions. Direct ways of driving specific actions include providing incentives such as subsidies for individuals to buy low-emission vehicles, while indirect methods commonly include measures like carbon taxes on emissions or carbon markets (see **Chapter 6**). However, given the enormity of the issue of addressing our reliance on the fossil fuels used throughout modern economies, and the need for an even speedier response from large scale actors such as companies, only affecting individual consumer choices will not be sufficient. In fact, the IPCC SR15 Report estimated that limiting the temperature increase to 1.5°C would require complete decarbonisation of the global energy system by 2050 (IPCC, 2018). In that context, whereas my individual actions may have negligible effects on reducing global emissions, the decisions and activities of big corporations— especially those in key energy-intensive economic sectors—would have pronounced effects in reducing emissions at the speed and scale required.

Breaking up (with fossil fuels) is hard to do, but a proactive state can get us through

A report published in 2017 estimated that just 100 companies in the fossil fuel industry (and its products) were responsible for 71% of all industrial GHG emissions produced since 1998 (Griffin, 2017). Even more pertinent, only 25 of those companies produce slightly over half of all global industrial emissions. Huge emissions reductions could therefore be achieved by affecting the calculation of costs and benefits made by those 100 companies, or even by only the 25 most polluting of them. Although these are some of the most successful and influential companies in the world, only 9% of them are owned by private investors. The remaining 91% are either public-investor owned (32%) or directly state-owned (59%) emissions producers (Griffin, 2017). This preponderance of state ownership of highly polluting companies means that states can play a crucial role in reducing GHG emissions through reducing the emissions from state-owned polluters.

Since only states have authority over the companies registered within their territorial boundaries, governments must intervene to change the calculation of costs and benefits that make mitigation actions by non-state actors possible and favourable. States can do this by directly or indirectly intervening in the economy and by creating policy certainty for non-state actors. For example, a study by the USA's Secretary of Energy Advisory Board identified six industries which could potentially produce emissions reductions in the gigatonne range (cf. global carbon emissions in 2020 are about 45 gigatonnes) if they were forced to internalise the costs they have thus far successfully externalised (Majumdar and Deutch, 2018). These six industries are agriculture and the oil, gas, coal, steel and concrete industries.

System level domestic interventions required

States are usually responsible for the health of their citizens and one of the key aspects of this is the provision of adequate nutrition. Successful reduction in GHG emissions from food production requires a system-level response that states are best placed to implement or facilitate. This is because the climate mitigation potential through the food system is predominantly (though not exclusively) found on the demand side—i.e. through changing consumer behaviour in favour of lower-carbon food consumption: less red meat, less over-all consumption, less food waste, and so on (Climate Action Tracker, 2018). States would need to attempt changes to the wider people-health-food system with efforts touching on multiple points of public policy— from nutrition education programmes, to labelling requirements, health and safety legislation, incentivising illness prevention instead of symptom-treatment in the health insurance industry, right the way through to possible sugar taxes, subsidising low-GHG products and perhaps even carbon taxes on meat production and consumption. Non-state actors, though they may lobby for change, just do not have the authority to implement the wide-ranging changes required.

As shown above, energy is crucial to almost every aspect of modern living and most of that energy has been (and is still) provided by the oil, coal and natural gas industries (in 2020, over 80% of total global energy consumption was derived from fossil fuels). In order to reduce emissions it is necessary both to rapidly increase the supply of zero-carbon electricity *and* to shift as much energy consumption as possible to use that electricity. In contrast to decarbonising electricity, supplying zero-carbon liquid fuels (e.g. for the transportation sector) is much more technically complex and not yet economically feasible at scale. A shift to zero-carbon electricity facilitates the possible reduction in emissions in a range of industrial sectors (steel manufacture, buildings, transport) where it is otherwise challenging to reduce the energy intensity of the product or process (Climate Action Tracker, 2017). Increasing the proportion of zero-carbon electricity requires that states stop subsidising fossil fuel producing entities, whether they own them or not. Again, this is something that non-state actors just cannot do.

In 2017 an International Monetary Fund working paper estimated that worldwide, coal and petroleum account for 85% of energy subsidies (Coady et al., 2019). Providing fossil fuel production subsidies effectively incentivises extraction by making fossil fuels cheaper in comparison to renewable energy sources and skews the market to make marginal fossil fuels profitable (Gerasimchuk et al., 2017). Broadly speaking, states are best placed to introduce energy pricing reforms to incorporate the environmental and social costs their use creates into the price of fossil fuels through the social cost of carbon (see **Chapter 5**). In absolute terms, of the top five states or confederations which subsidise fossil fuels—China, the USA, Russia, the EU and India (Coady et al., 2019)—four are also consistently the top global CO_2 emitters: China, the USA, Russia and India. It is clear that states remain crucial to successful emissions reduction. States need to assist renewable energy providers (through tax rebates, production subsidies, support for innovation and so on) and also incentivise the development and consumption of zero-carbon electricity and associated products (see **Chapter 7**).

Policy certainty allows non-state actors to make plans

Technologies can take years to research, develop, test and bring to market. Certainty about the future domestic policy environment within which non-state actors operate and to which they need to comply enables them to take the risk of investing in long-term technology development. Policy certainty also allows non-state actors time to plan and respond to legislation or regulations. States can, for example, require companies to formulate and communicate low carbon transitions plans or to invest in R&D to decarbonise their products or processes. It is only within the state's purview to provide a legislative framework within which the public and private sector must operate. Non-state actors cannot do this. An example of a legislative framework is the UK's Climate Change Act of 2008 which signalled a long-term emissions trajectory for the country. The 2008 Act committed the UK to reducing its carbon emissions to 80% below the 1990 baseline by 2050, and in June 2019 the UK Parliament amended this by law to make the target 'net zero' emissions by 2050. Non-state actors can now begin to make decisions in line with that target. They know that they will not be unilaterally incurring costs when researching, developing and implementing changes towards low-emission technologies (for instance), because all other non-state actors are subject to the same strictures imposed by the law.

Some state actors also have at their disposal significant R&D funding. Historically, states have deployed their R&D funds in areas in which there may be a good outcome for the public at large, but where there is no apparent incentive for private finance (initially at least) to invest. States have also funded R&D in technology areas where there is a high risk of failure or no apparently marketable outcome. A notable example of state-sponsored technology which revolutionised the world was the development of the internet (initially called ARPANET) developed in the late 1960s by the US Defence Department's Advanced Research Projects Agency (also see Sarewitz, 2014). States could be active in earmarking R&D funds for research that identifies how these major polluting industries might decarbonise or reduce the energy intensity of their products and operations. What makes such public funding different from private funding is the possibility that it could be part of a well-designed industrial innovation strategy implemented by the state. This would create system-level changes in industrial production methods and accelerate reductions in GHG emissions.

Conclusion

Humanity's continued system-level, institutionalised reliance on fossil fuels to underpin economic life and social welfare is causing many of the changes to the climate and biosphere we are seeing around the world. Ultimately, the severity of the social, environmental and economic impacts of climate change will be determined by the length of time taken to reach net zero carbon emissions. Most states have *not* acted swiftly to address the global challenges identified in successive IPCC reports. This is part of the reason why the problem of climate change now requires such sweeping changes to how we produce and consume energy that *only* states can now facilitate the kinds of broad-ranging, systemic-level changes that are needed. In order to alter the calculation of costs and benefits of engaging in some kinds of behaviour, an overarching authority—like a state—must implement incentives, regulations and policies that internalise the negative externalities to reflect the true cost or consequences of the use of a shared or common resource.

Although non-state actors have a role to play, they cannot provide the necessary enabling or constraining national policy environments. Successful emissions reductions lie more in the hands of state than non-state actors.

Further reading

Andonova, L.B., Hale, T. and Roger, C. eds. (2018) *The Comparative Politics of Transnational Climate Governance*. London: Routledge.

This edited collection of essays explores how domestic political, economic and social forces systematically shape patterns of non-state actor participation in transnational climate initiatives. The contributing chapters explore the role of cities, non-governmental organisations, companies, carbon markets and regulations, as well as broader questions of effectiveness and global governance.

Bulkeley, H. (2015) *Accomplishing Climate Governance*. Cambridge: Cambridge University Press.

Through a range of case studies drawn from communities, corporations and local government, Harriet Bulkeley examines how climate change comes to be governed and made to matter as an issue with which diverse publics should be concerned. Rather than seeking the solution to climate change 'once and for all', she argues to ameliorate the climate problem through pursuing more progressive ends.

The Climate Action Tracker (https://climateactiontracker.org)

The Climate Action Tracker (CAT) is an independent scientific analysis produced by three research organisations tracking climate action since 2009. This website quantifies and evaluates climate change mitigation commitments and assesses whether countries are on track to meet them. It aggregates country commitments to the global level and estimates the likely range of global warming by the end of the century.

Climate One podcasts (https://climateone.org/watch-and-listen/podcasts)

These podcasts from Climate One at The Commonwealth Club offer a variety of thought pieces from top thinkers and doers worldwide from business, government, academia and advocacy groups. They are concerned with questions around climate, energy, environment and governance.

Jordan, A., Huitema, D., Van Asselt, H. and Forster, J. eds. (2018) *Governing Climate Change: Polycentricity in Action?* Cambridge: Cambridge University Press.

This book brings together contributions from some of the world's foremost environmental policy experts to provide the first systematic test of the ability of polycentric thinking, as opposed to state-centric governance, to explain and enhance societal attempts to govern climate change.

Follow-up questions for use in student classes

1. Why is it possible to see climate change as a complex, multi-scale and distributed issue? What are the potential implications of this insight for climate mitigation?
2. How might state and non-state actors best work together to bring about emissions reductions? How is the agency behind non-state climate action different from that of states?
3. What are some of the sector- and place-specific incentives that might drive non-state engagement in climate mitigation and adaptation? Could you provide a specific example?
4. What different policy tools or implementation strategies do different forms or types of states (e.g. democratic, theocratic, communist) have at their disposal?
5. Are there risks associated with increasing the engagement of non-state actors in climate mitigation? What are the best ways to anticipate and reduce such risks?

Notes

1 See http://web.unep.org/ourplanet/september-2015/unep-publications/climate-commitments-subna tional-actors-and-business-quantitative. Accessed 22 July 2019.
2 See www.irena.org/publications/2018/Jan/Global-Landscape-of-Renewable-Energy-Finance. Accessed 22 July 2019.
3 A brief overview of the arguments can be found here: www.globalpolicyjournal.com/blog/26/04/ 2016/paris-agreement-when-treaty-not-treaty. Accessed 21 May 2019.

References

Anadon, L.D., Chan, G., Harley, A.G., Matus, K., Moon, S., Murthy, S.L. and Clark, W.C. (2016) Making innovation work for sustainable development. *Proceedings of the National Academy of Sciences.* **113**(35): 9682–9690.

Andonova, L.B. (2017) *Governance Entrepreneurs: International Organizations and the Rise of Global Public Private Partnerships.* Cambridge: Cambridge University Press.

Andonova, L.B., Hale, T. and Roger, C. (2017) National policy and transnational governance of climate change: substitutes or complements? *International Studies Quarterly.* **61**(2): 253–268.

Bernstein, S. and Hoffmann, M. (2018) The politics of decarbonization and the catalytic impact of subnational climate experiments. *Policy Sciences*. **51**(2): 189–211.

Bettencourt, L.M.A., Trancik, J.E. and Kaur, J. (2013) Determinants of the pace of global innovation in energy technologies. *PLOS ONE*. **8**(10): e67864.

Bulkeley, H., Andonova, L.B., Betsill, M.M., Compagnon, D., Hale, T., Hoffmann, M.J., Newell, P., Paterson, M., VanDeveer, S.D. and Roger, C. (2014) *Transnational Climate Change Governance*. Cambridge: Cambridge University Press.

Climate Action Tracker. (2017) Foot off the gas. Increased reliance on Natural Gas in the power sector risks an emissions lock-in [online]. *Climate Action Tracker*. Accessed 19 July 2019. Available at: https://climateactiontracker.org/publications/decarbonisation-memo-series/

Climate Action Tracker. (2018) What's on the table? Mitigating agricultural emissions while achieving food security [online]. *Climate Action Tracker*. Accessed 19 July 2019. Available at: https://climateactiontracker.org/publications/decarbonisation-memo-series/

Coady, D., Parry, I., Le, N-P. and Shang, B. (2019) *Global Fossil Fuel Subsidies Remain Large: An Update Based on Country-Level Estimates*. WP/19/89. Geneva, Switzerland: IMF.

Gerasimchuk, I., Bassi, A.M., Ordonez, C.D., Doukas, A., Merrill, L. and Whitley, S. (2017) *Zombie Energy: Climate Benefits of Ending Subsidies to Fossil Fuel Production*. Working Paper. Winnipeg, Canada: International Institute for Sustainable Development.

Green, J. (2014) *Rethinking Private Authority Agents and Entrepreneurs in Global Environmental Governance*. Princeton, NJ: Princeton University Press.

Griffin, D.P. (2017) *CDP Carbon Majors Report 2017*. London: Carbon Disclosure Project, Climate Accountability Institute. Accessed 19 July 2019. Available at: https://cdp.net/en/articles/media/new-report-shows-just-100-companies-are-source-of-over-70-of-emissions

Hale, T. (2016) All hands on deck: the Paris Agreement and non-state climate action. *Global Environmental Politics*. **16**(3): 12–22.

Hoffmann, M.J. (2011) *Climate Governance at the Crossroads*. Oxford: Oxford University Press.

IPCC (2018) Summary for Policymakers. In: *Global Warming of 1.5°C. An IPCC Special Report on the impacts of global warming of 1.5°C above pre-industrial levels and related global greenhouse gas emission pathways, in the context of strengthening the global response to the threat of climate change, sustainable development, and efforts to eradicate poverty* [Masson-Delmotte, V., P. Zhai, H.-O. Pörtner, D. Roberts, J. Skea, P.R. Shukla, A. Pirani, W. Moufouma-Okia, C. Péan, R. Pidcock, S. Connors, J.B.R. Matthews, Y. Chen, X. Zhou, M.I. Gomis, E. Lonnoy, T. Maycock, M. Tignor and T. Waterfield (eds.)]. World Meteorological Organization, Geneva, Switzerland, 32 pp.

Keohane, R.O. and Victor, D.G. (2011) The regime complex for climate change. *Perspectives on Politics*. **9**(1): 7–23.

Majumdar, A. and Deutch, J. (2018) Research opportunities for CO_2 utilization and negative emissions at the Gigatonne scale. *Joule*. **2**(5): 805–809.

Michaelowa, K. and Michaelowa, A. (2017) Transnational climate governance initiatives: designed for effective climate change mitigation? *International Interactions*. **43**(1): 129–155.

Ostrom, E. (2014) A polycentric approach for coping with climate change. *Annals of Economics and Finance*. **15**(1): 97–134.

Sarewitz, D., ed. (2014) *The Rightful Place of Science: Government and Energy Innovation*. The Consortium for Science, Policy and Outcomes. Tempe AZ: Arizona State University.

Sinnott-Armstrong, W. (2005) It's not *my* fault: global warming and individual moral obligations. In W. Sinnott-Armstrong and R.B. Howarth, eds. *Perspectives on Climate Change: Science, Economics, Politics, Ethics*. Bingley, UK: Emerald Group Publishing Limited. pp. 285–307.

13 Is legal adjudication essential for enforcing ambitious climate change policies?

Eloise Scotford, Marjan Peeters and Ellen Vos

Summary of the debate

This debate considers what is the appropriate role for the courts as an actor in climate politics. **Eloise Scotford** argues that legal adjudication is inevitable and essential in enforcing ambitious climate change policies. She argues that courts have an essential role in ensuring that climate policies are delivered in a fair and legally accountable way, in deciding inevitable legal disputes and in expressing symbolic community statements relating to climate change. In contrast, **Marjan Peeters** and **Ellen Vos** contend that given the complexity and uncertainty in climate science, and the many interests affected by emissions reduction policies, courts are not equipped or entitled to replace political decision-making. The level of emissions reductions in a specific country, and the pathways for achieving those reductions, are primarily political questions and not legal ones.

YES: Because of the inevitability and necessity of legal disputes about climate change (*Eloise Scotford*)

Introduction

Legal adjudication is essential for enforcing ambitious climate policies, for two reasons. First, national courts have an inevitable role in our current international legal regime for responding to climate change. Second, the roles of courts and adjudication in legal systems represent essential functions in implementing and enforcing ambitious climate policies. The inevitable role of courts in enforcing climate change policies, particularly national courts, is driven by the nature of climate change and by the task now legally mandated to tackle this widespread social and environmental challenge through international law. Being a '**wicked problem**', climate change is inherently disruptive of regulatory and legal systems, giving the courts an inevitable role in adjudicating new legal questions and emerging social tensions that manifest through legal disputes (Fisher et al., 2017). Further, the **Paris Agreement** puts legal action at the national scale squarely in the frame of climate policy. This is the scale at which the jurisdiction of courts—and access to them—is more routine, heightening the role for national courts in delivering ambitious climate policy.

Not only is litigation concerning climate change inevitable, but the roles played by courts in adjudicating disputes relating to climate policy are fundamental. First, at least in constitutional democracies, courts have a fundamental constitutional role in holding governments to account. Thus, when governments commit to national climate policies—in the form of **nationally determined contributions** (NDCs) under the Paris Agreement, or otherwise such as through national climate policies or statutes—national courts will play a role in ensuring governments honour their commitments. This is enhanced by the structure of the Paris Agreement, which shifts accountability for parties' climate policy commitments from international mechanisms (as occurred under the **Kyoto Protocol**) to the national level. Adjudicative processes themselves perform important roles in the climate change context. Courts must apply legal doctrines to legal disputes relating to climate policy that inevitably arise (their *dispute resolution* function), particularly where legal doctrines need to adapt or adjust to novel legal questions. Courts also give authoritative statements about those conflicts to the wider community (their *symbolic* function) and provide a lens for viewing the intricate social and economic implications of climate policy.

Two caveats or clarifications are important at the outset. First, there is no single type of legal dispute or court case relating to climate change. Cases can range from those that challenge national government climate policy generally, to those that concern the legal nature of rights created by regulatory schemes (such as an emissions trading scheme; see **Chapter 6**) or the determination of legal obligations when private individuals make bargains or arrangements against the backdrop of climate policy. This essay defines 'climate change cases' broadly, covering any national adjudicative processes that relate to climate policy in some way. These cases will vary in subject matter and legal form. They will be covered by procedural and jurisdictional rules, and constitutional constraints, that are specific to the jurisdiction involved. Second, in focusing on adjudication, not litigation, the chapter's argument is focused on the role of courts and other adjudicative bodies (such as tribunals) rather than litigation strategies that interest groups or parties may seek to pursue. There is a thriving community of those involved in strategic climate litigation, but this is dependent on the fundamental roles of courts and tribunals that apply and determine the law within legal systems.

The 'legally disruptive nature' of climate change: climate adjudication is inevitable

Climate change is legally disruptive. In a previously co-authored paper (Fisher et al., 2017), we outlined four characteristics of climate change that make it disruptive in a legal sense:

- its causes and consequences are polycentric;
- the trajectory of climate change is scientifically uncertain;
- it inherently gives rise to sociopolitical conflict (both as a phenomenon and through ambitious policy efforts to address climate change);
- it is a physically dynamic phenomenon.

Each of these aspects of climate change creates difficulties for legal processes and doctrines that assume a stable natural environment and relatively stable relational and political structures in relation to which legal rights and obligations operate. They also mean that legal disputes concerning the 'wicked' social and environmental phenomenon of climate change are inevitable. The law that applies to such disputes will not always be obvious, making adjudicative processes to determine the relevant law and to settle the positions of conflicting parties particularly important.

To exemplify this inevitability, let us consider ambitious climate policies that incentivise a transition towards low carbon industries of energy production. In any such energy transition, there will be winners and losers and government policies will manage these changes more or less well. If poorly managed, legal disputes will arise and are proper avenues for those with grievances. Those industries forgoing revenue-making opportunities will have economic grievances and may seek legal recourse for impaired legal rights (e.g. they may bring claims for breach of contract, infringed constitutional rights, or review of government action on grounds of legitimate expectations or other public law grounds). New industry players may also have economic grievances that manifest in legal disputes (as in the case of government feed-in tariffs that are quickly withdrawn due to oversupply). There might also be disappointed workers in high carbon emitting industries who lose jobs, or else citizens for whom energy costs rise quickly in an 'unjust' transition. Areas of law such as contract law, labour law and public law contain doctrines that will often be relevant in these kinds of circumstances. Even if well managed, major structural transition brings economic and social strain and legal arguments will arise. The economic, social and legal status quo is being unsettled by climate change and legal claims express that disruption. These arguments must be settled for climate policy to be properly implemented.

As another example, let us say that a national government has adopted an ambitious national climate policy to achieve **net zero emissions** of GHG gases by 2035, setting out this mitigation goal in a widely publicised government policy document. There is, however, no 'climate change department' in this country's government and existing government departments continue working within their siloed portfolios. In this particular country, the Department for Transport has been working on airport expansion plans for ten years and determines that the issue of inadequate airport capacity is now urgent to resolve, recommending that the country's biggest airport be expanded. In its view, the government's flagship climate change commitment is not a barrier to this proposed airport expansion, because it does not require that any specific policy or development must change. In the department's view, climate policy innovation can come from somewhere else—after all, this new airport expansion policy is in the public interest.

Here we see a direct policy clash between climate and transport policy. This policy mismatch within the existing structures of government can be argued about politically but, ultimately, there is a question of accountability. The courts are the central organ of legal accountability in many jurisdictions and political systems globally. In this example they would provide a key forum in which non-political actors could ask the government whether its airport expansion policy is rational. Such judicial challenges to proposed airport expansions have, inevitably, been instigated (e.g. in the UK regarding Heathrow[1] and in Ireland regarding Dublin's airport[2]).

These two examples each highlight how the radical change required by ambitious climate policies makes legal disputes inevitable. The large number of climate-related cases now being heard in courts around the world bears this out (Fisher et al., 2017). These examples show how such disputes are essential—both in determining where rights and liabilities lie between private citizens and in settling where public lines of accountability lie in governments.

The Paris Agreement: structuring national climate adjudication

This inevitable pressure on national courts is enhanced by the structure of the Paris Agreement 2015. With its hybrid legal architecture—an international treaty relying on and constructed by nationally driven action—the Paris Agreement's obligations are to be determined *and* implemented at the national scale. Granted, there are international accountability mechanisms within the Agreement. In particular, there is Article 13's non-punitive transparency mechanism to account for national action and support measures, subject to technical expert review, and Article 15's non-punitive facilitative compliance mechanism involving the establishment and operation of an expert committee. But there is no formal enforcement mechanism at the international level for the national commitments made by signatory countries under the Agreement. Domestic legal action fills this compliance void and is a critical forum for holding governments to account for their commitments under the Agreement. This is supported by Article 4(2), which provides that 'Parties *shall* pursue domestic mitigation measures, with the aim of achieving the objectives [of successive NDCs]' (UNFCCC, 2015). This is a mandatory requirement (not all obligations in the Agreement are expressed in such mandatory language), indicating that legal enforcement is appropriate and necessary and must be situated at some level in the Agreement's hybrid governance structure.

In this vein, there has been a rise in the number of cases in different jurisdictions around the world since the 2016 ratification of the Paris Agreement, with various NGOs or groups bringing public law claims (tailored to jurisdictional contexts) against national governments in Party countries. These claims challenge various aspects of national government climate policy for not being sufficiently ambitious when measured against the goals of the Agreement and/or NDCs. At the time of writing, cases of this nature had been launched at least in the courts of the European Union, Germany, Ireland and the UK.

The UK case in *R (Plan B and [others]) v Secretary of State for Business, Energy and Industrial Strategy* (2018)[3] demonstrates how succeeding in these cases can be difficult. Nevertheless, their agitating and publicising impact can be considerable, providing a mechanism for indirect enforcement. In *Plan B*, the claimant NGO argued that the Secretary of State's refusal in light of the Paris Agreement to revise the UK's 2050 carbon mitigation target—set in the Climate Change Act 2008 to require reduction of GHG emissions by 80% from the 1990 level—meant that the UK Government was in breach of its international law obligations. The case was ultimately unsuccessful on the grounds that the government had not behaved irrationally and had a wide discretion in determining its climate policy. However, in the meantime the government referred a question to the national Committee on Climate Change asking it to reassess the UK's 2050 emissions target. The committee recommended that the target should indeed be revised to a 'net zero' emissions target and, in June 2019, Prime Minister Theresa May announced that legislation had been laid before Parliament to revise the statutory target accordingly.

The role of courts: holding national governments to account for climate policy

Even beyond the governance structure of the Paris Agreement, national courts play a fundamental role in holding their governments to account for executing the laws that have been enacted and in applying their own policies properly. Again, the constitutional and administrative law (that is, public law) frameworks for these legal accountabilities will vary in different jurisdictions. Some will have a formal written constitution against which public actions can be tested. Some will have public law doctrines developed over time and applied by the courts—whether based in ideas of reasonableness, relevance of considerations taken into account, rationality, legitimate expectations, procedural fairness, proportionality or other norms of good administration. And some will have codes or statutes that set out public law norms, which may include human rights (see **Chapter 11**). All these public law avenues are legal doctrines that embody and apply ideals of the rule of law, ensuring that those who govern us do so lawfully. The courts play an essential constitutional role in upholding these norms.

Thus, again, it is unsurprising that public law actions of various forms have been brought to hold governments to account for their existing climate policies or for pushing them to adopt more ambitious climate policies. Whether or not these actions are successful, this adjudicative function is essential for upholding the rule of law and good governance. And some actions have been successful. The Pakistani case of *Ashgar Leghari v Federation of Pakistan* (2015)[4] is a well-known case in which the Pakistan Government was found to have been taking inadequate climate action on human rights grounds. The domestic Pakistan court issued quite interventionist remedies requiring the government to improve its administrative arrangements for implementing its own climate policy.

Another very well-known case brought against a government challenging its climate policy is *Urgenda Foundation v The Netherlands* (2015).[5] Many different legal arguments were made in this case brought against the Dutch Government by an environmental NGO—arguments based in human rights, in the civil law doctrine of 'hazardous state negligence' and in EU and international law. Most of these arguments had legally novel aspects, whether on issues of legal procedure or liability. In both the lower District Court and on appeal in the Hague Court of Appeal, the State of the Netherlands was found to have adopted an unlawful national climate policy that was insufficiently ambitious in the short-term. In response to arguments that the court had overstepped its constitutional role in dictating Government policy in deciding this case, the Court of Appeal was clear that the state is not above the law. It needs to be held to account for any unlawful actions, particularly those involving breaches of human rights.[6]

The role of adjudication: dispute resolution and symbolism

In thinking about legal adjudication and climate change it is easy to fall into the trap of supposing that adjudication is simply an instrumental route for compensation and justice. However, that is a caricature of law and of the role played by adjudication in courts and similar fora (although compensation and justice are, of course, important). Legal adjudication plays multiple roles, including (1) applying applicable laws and doctrines to legal disputes; and (2) performing a powerful symbolic function for communities bound by the rule of law (Fisher et al., 2017). In relation to both of these functions, legal adjudication is inevitable and essential in enforcing ambitious climate change policies.

In terms of their *dispute resolution function*, courts are settling what the law is, whether in public law cases (as explained above) or in private law cases (where individuals, either people or companies, sue each other for some form of civil liability). This task can be straightforward, but often involves novel legal reasoning due to the disruptive nature of climate change (Fisher et al., 2017). Thus, in *Urgenda*, as indicated above, different legal arguments were successful at the different stages of the litigation, involving novel legal reasoning. For example, it was not clear whether the Urgenda Foundation (representing Dutch citizens, but also citizens from other countries and future generations) should have standing to bring the legal claim in court. Standing doctrines are common issues of legal procedure, but there were novel facts to deal with in this climate context.

Similarly, arguments of causation were novel. Why was the Dutch Government being held accountable for *its* contribution to global climate change, which was relatively small and ultimately unconnected to any particular environmental harm? The court adapted the relevant test of causation to apply to this wicked environmental problem, acknowledging that applying existing tests of causation might mean that no country was responsible for climate change and its harmful consequences ('an effective legal remedy for a global problem as complex as this one would be lacking'; *Urgenda*, The Hague Court of Appeal). Whether courts will always adapt their legal procedures and doctrines to accommodate climate change in this way is subject to many factors and constraints of national legal culture, but ultimately a court needs to resolve the legal issues before it.

Resolving legal disputes between parties not only involves deciding and applying the law; it also sheds light on aspects of climate policy that might not otherwise be visible. The wide range of (sometimes quite technical) legal claims relating to climate policy demonstrates how the polycentric causes and implications of climate policy reach into very widespread spheres of social and economic life (see **Chapter 12**). For that reason again, climate adjudication—the settling of norms and rules in the intricate, everyday and widespread dealings of our societies and economies—is vital for delivering climate policy in a deep and socially transformative way (Bouwer, 2018).

The second way in which climate adjudication is vital for enforcing ambitious climate policy is through its *symbolic function*. In articulating disputes concerning climate policy, in hearing them within the established and respected processes of the courts, in finding facts and in establishing legal liability, courts send messages to their communities (local, national, global) in the cases that they hear and decide. In cases concerning climate change, adjudication often has huge social significance and resonance (Fisher et al., 2017). This can be demonstrated by the 2007 US case of *Massachusetts v EPA*,[7] which essentially concerned an issue of statutory interpretation of the US Clean Air Act (whether air pollutants included GHGs for the purposes of pollution regulation of vehicles). In the course of deciding the case, the Supreme Court stated that '[t]he harms associated with climate change are serious and well recognized'. The impact of this statement in a judgment of the US Supreme Court had huge symbolic significance, quite apart from the legal issue resolved in the case (Jasanoff, 2015). Climate change was a fact, and a serious issue, so said the court.

Similarly, the 2019 New South Wales Land & Environment Court case of *Gloucester Resources v Minister for Planning*[8]—a planning appeal confirming that permission should not be granted to construct an open cut coal mine—had huge symbolic value, resonating globally as a ground-breaking climate case. In fact, resolving the planning appeal in the reasoning of the case did not turn on the climate impacts of the proposed mine, but the judge's careful consideration of climate change issues in the judgment was nonetheless seen more widely as very significant.

Resolving climate-related legal disputes between parties also serves a symbolic function in a regulatory sense, due to the signalling function of decided cases. In private law claims in particular—for example, in contract, tort, or company law—the outcomes of these cases send signals about how businesses, individuals and other actors should be acting in relation to climate change. For example, in company law and pensions law, claims might be made about whether companies or pension funds have properly taken into account climate change factors in applying the requirements of corporate reporting or pension investing. At the time of writing, these kinds of private law claims were being brought in Australia and being considered elsewhere. The findings in such cases settle claims brought by individual shareholders or investors, but they also send a much wider message about how companies and pensions funds *should* operate in relation to climate change issues. These individualistic private law claims can provide a 'regulatory pathway' to delivering ambitious climate policy (Peel and Osofsky, 2015) and thus are an essential aspect of enforcing ambitious climate policies.

Conclusion

The essential and inevitable nature of adjudication in enforcing ambitious climate policies does not mean that court actions or judgments alone will deliver ambitious climate policies. Far from it. It is fundamentally the architecture of a state's government and its administration that will deliver climate policies (see **Chapter 12**). However, courts have a critical role on holding that public action to account and in confirming and expressing the obligations of myriad actors (public and private) in relation to the widespread social transition required by ambitious climate policies.

There is a risk that climate adjudication is misunderstood or that it becomes the object of too many political hopes. Court cases are often seen as either 'wins' or 'losses' for climate policy. Judges can incorrectly be seen as saviours or untamed activists in the climate cause, as hinted at below by **Marjan Peeters** and **Ellen Vos**. These are misperceptions of the roles of courts and adjudication. This is not to deny that courts face challenges in applying legal principles and doctrines to disputes involving climate change, which are often legally disruptive due to the polycentric, socially contested, scientifically uncertain and dynamic nature of climate change. But the proper and fundamental roles of the courts should always be kept in mind. Courts must decide cases that come before them. They must declare what the law is. And they must hold governments to account. In doing so, they are often deciding very ordinary issues of legal doctrine or interpretation, albeit in a very complex context which can make reasoning novel and challenging. Their job in adjudicating climate disputes is all the more important for that. Legal issues must be settled if climate policy is to be enforced and, in the process, its full implications are rendered visible.

In the wake of ambitious climate policy, climate adjudication is inevitable and essential in societies subject to the rule of law.

NO: Judges should remain judges and should not become scientists or policymakers (*Marjan Peeters and Ellen Vos*)

Introduction

Enormous societal and technological transformations are needed at all levels to address the challenges of climate change: at local, national, European and international scales. To this end, the international climate treaty regime provides a legal framework for cooperation among states in order to achieve the aim of holding the increase of global average temperature well below 2°C and to pursue efforts to limit this to 1.5°C (as stated in article 2 of the Paris Agreement; UNFCCC, 2015). Scientific insights provide support for this international policy ambition codified in treaty law.[9] However, the Paris Agreement does not delineate what emissions reduction specific countries have to undertake and the role of courts in this undertaking is therefore limited.

Certainly, courts play an essential role in holding governments to account, by testing the legality of their decision-making. Next to this, courts can also determine liability of natural and legal persons when they cause damage to others. However, courts can also be called upon to declare as 'unlawful' governmental emissions reduction targets that are not the most ambitious for protecting the climate system or to order governments to adopt a certain emissions reduction policy. This raises a fundamental question about whether courts are the right institutions to take these kinds of political decisions.

Core arguments will be presented in this essay that explain why courts do, and should, face limits when adjudicating in the field of climate change, including human rights adjudication. More particularly, we will show why courts are not in the best position to decide on questions relating to the *rate* at which a country should reduce its total emissions, nor *how* a certain volume of emissions reduction should be realised. Before proceeding, we need to point out important methodological restrictions in our argument. Around the world, national jurisdictions vary regarding the role of the courts, including the question of whether public interest cases, such as the importance of avoiding dangerous climate change, can legitimately be brought to the courts (see for example the case of Switzerland: Bähr et al., 2018). Moreover, judicial adjudication largely depends on the specific circumstances and strategy of a claim. The arguments offered here find their roots in the notion of the rule of law on the European continent and, more specifically, within the European Union and its Member States.[10]

Judicial review

Judicial review of governmental decisions, including legislation, is an important aspect of a legal system operating under the rule of law (Türk, 2009: 1). Access to courts is necessary to ensure that governments respect human rights—such as the rights codified in the Charter of Fundamental Rights of the European Union (European Union, 2000)—and that they adhere to the often more specific rules, such as environmental laws, adopted through democratic processes. However, courts do not have the task to fill all the gaps that democratically accountable decision-makers leave open. Not all societal issues are 'justiciable'. This means that if authorities take debatable decisions—in the sense of not providing the highest level of welfare possible to society (if that level can at all be determined)—focus should be placed on how to improve decision-making in the legislative and the executive branches of government, instead of immediately reverting to a court to overturn that decision.

Political decisions on how to address climate change

The main issues that need to be decided in order to address climate change are of a political nature. For instance, the IPCC lists various portfolios of mitigation measures for achieving the 1.5°C warming target, thereby 'striking different balances between lowering energy and resource intensity, rate of decarbonisation, and the reliance on carbon dioxide removal' (IPCC, 2018: 12). This implies that important policy choices have to be made when determining the direction a specific country will take for combating climate change. In the course of achieving this direction, optimal solutions for each specific country need to be determined, for example the roles of nuclear energy, carbon capture and storage and/or renewables (see **Chapter 7**). Given the available options, and their relative impacts on the environment, economy and society, it is an explicitly political decision as to the total emissions reduction level a country will adhere to by a given date. If judges would engage in such decision-making, they could appear to be the primary standard setters in society. This is not in line with case law of the Court of Justice of the European Union (CJEU), which gives quite some leeway to political decision-making in complex socioeconomic matters, including in the environmental area. Judicial practice shows that the CJEU will only intervene in case of a manifest error of appraisal by the authorities, particularly by the EU legislator.

The following two examples illustrate that this practice is also evident in the field of EU climate law, more particularly by means of a rather light judicial review of legislation adopted by the EU institutions. In a seminal case decided in 1998, an appellant *inter alia* argued that a regulation on substances that deplete the stratospheric ozone layer had too narrow a scope because it did not take into account the Global Warming Potential (GWP) of the substances involved. The CJEU only investigated whether there was a manifest error of appraisal and concluded that this was not the case (C-341/95, Bettati v. Safety HiTech, para 35, 53).[11]

In another core EU climate law case, decided in 2008, the CJEU endorsed a step-by-step approach taken by the EU legislator, which meant that the legislator was not obligated to include all relevant GHG emitters at once in its legislation. Basically, the EU legislator was granted 'a broad discretion where its action involves political, economic and social choices and where it is called on to undertake complex assessments and evaluation' (C-127/07, Arcelor para 57).[12] This does not imply that the court does not assess at all the decision-making of the governmental legislator, but merely that it checks whether the choice of the legislator is based on 'objective criteria appropriate to the aim pursued by the legislation in question' and whether the legislator took 'fully take into account all the interests involved' (Arcelor para 58–59). Nonetheless, when carrying out its assessment, the CJEU defers largely to the legislator by assessing that the exercise of its discretion 'must not produce results that are *manifestly less appropriate* than those that would be produced by other measures that were also suitable for those objectives' (Arcelor para 59, emphasis by authors).

Even Christopher Stone, who in 1972 already called for establishing a strong protection of the environment before the courts in his seminal publication 'Should trees have standing?', points at the limits of the courts in regard to climate change. According to Stone, the political question of how much to reduce GHGs is ill-suited for courts to adjudicate (Stone, 2010: 34, 53). Furthermore, there is even concern that in judicial decision-making on rather open-ended issues personal beliefs may play a role in developing the outcome of a case. (In practice this is hard to determine, since judges will be unlikely to reveal such influence; see Bergkamp and Hanekamp, 2015; Petersen, 2017: 359–360.)

Nonetheless, claims aiming to achieve 'more political action' on climate change are now widely submitted before courts around the world. Case law will show where, in various jurisdictions, the line will be drawn between the political and judicial sphere. Peculiarly, it is the courts themselves who need to draw this line. Yet if courts move in such a direction that politicians (or society at large) would judge as over-stepping their perceived role, this may result in governments either neglecting court decisions or else taking measures to restrict the powers of the court (for examples at the international level see Petersen, 2017: 364). In summary, when courts become dominant in the sense of taking the lead with regard to the content of state policymaking, states may try to find ways to escape or mitigate this power.

Moreover, at the international level there is no clear legal criterion that determines what emissions reduction effort a country has to deliver. The Paris Agreement (see article 4) obliges parties to communicate their nationally determined contributions (NDCs) to emissions reduction. It states that developed countries should continue to take the lead by undertaking economy-wide absolute emissions reduction targets. If parties want to escape from this agreed effort, they can seek to leave the Paris Agreement, as the USA has already notified to do. Also the International Court of Justice (ICJ), if it ever will be called upon in a climate case, needs to operate carefully, as noted here:

> [...] the potential benefits of relying on the rule of law must be balanced against the potential disruptions such an adjudicative approach could cause to the multilateral effort of a treaty-based solution that aims at addressing both the causes and the effects of climate change.
>
> (Voigt, 2016: 156; referring to Bodansky, 2016)

Practice shows that Japan and France have limited the jurisdiction of the ICJ following environmentally related decisions on, respectively, whale catchments and nuclear tests (Petersen, 2017: 364). Such boomerang effects—weakening the power of courts after taking, in the eyes of the condemned, a too interventionist role—are also not unimaginable for the EU. If the CJEU were to decide that the emissions reduction commitments of individual Member States as adopted by EU legislation are insufficiently ambitious, this could spur even more resistance within national orders against this EU judicial influence.[13]

Role of judges in cases of complexity and scientific uncertainty

Although climate science continues to progress, substantial uncertainties about the course of future climate remain (see **Chapter 2**). For instance, the way global temperature is measured affects the estimated remaining carbon budget and 'uncertainties in the size of these estimated remaining carbon budgets are substantial and depend on several factors' (IPCC, 2018: 14). While the **precautionary principle** legitimates environmental protection when the danger cannot be sufficiently proven, the precise intensity and form of governmental action for such protection needs to be determined within a sphere of uncertainty and by balancing different interests, including costs. The value and application of the precautionary principle is the subject of an intense scholarly debate. This ranges from support for a strict approach to the avoidance of risks, on the one hand, to warnings against over-regulation on the other (De Sadeleer, 2016). The statement that 'climate change today is not a matter of precaution, but one of prevention' stands in contrast with the uncertainties pointed out by the IPCC (Bähr et al., 2018: 208).

Parties to the Paris Agreement have committed to take climate action which they have to determine in their national plans. But the precise measures for each country have to be developed within a context of uncertainty. The consequences of specific policy choices— including the side effects on the economy, the environment and society—can be hard to predict. Such is the case, for example, with afforestation as a means to capture CO_2 from the atmosphere. While the Paris Agreement implies that climate action has to be taken by state governments (see **Chapter 12**), there is no single solution for how they should do so. Consequently, given the scientific uncertainties, but also the many interests at stake when deciding on (international, European and) national climate policies, judges are poorly equipped to assess the merits of a climate policy that is adopted through the democratic process.

When considering the best policy options in complex matters other mechanisms for evaluation and deliberation exist. These include, for example, the establishment of scientific or expert committees or agencies, or carrying out impact analysis of different proposed policy measures. Such procedures can inform governments on possible policy options—including their risks, although even the risks are not always known—which can then be debated through democratic process. Such mechanisms may not work perfectly and of course need to be critically scrutinised by independent scholars but, given the inevitable need for risk trade-offs, they would seem a better fit than judicial decision-making for the adoption of climate policies in democratic societies.

A further concern relates to the question of to what extent judges are capable of engaging with climate science in their decision-making. Studies have pointed to potential failures related to the understanding of the limits of science, including awareness that scientific consensus is a social construct (see **Chapter 9**) (Bergkamp and Hanekamp, 2015: 107). In legal practice, courts assess whether decision-makers are duly prepared, including in the use and interpretation of scientific studies. Courts are indeed important to guide the policy process, for instance by checking whether public participation in administrative decisions was secured (e.g. the Aarhus Convention, 1998) or whether there is appropriate disclosure of scientific advice and of impact analysis reports. Such a role for the courts does not mean, however, that the courts themselves will *impose* a certain reduction percentage in emissions on a government or order a certain policy measure to be taken.

Case law from the CJEU shows that, in general, the court increasingly enters into testing whether (administrative) decisions are properly based on relevant scientific information. They thereby function as an 'informational catalyst' (Scott and Sturm, 2007; see Vos, 2013: 144, for a rather positive appraisal of this development). This development increases the possibilities of the court to hold governments to account. Nonetheless, this judicial approach to information scrutiny needs to be further discussed by examining its application in the case law. A core question in this regard is whether the judges—who are educated as lawyers and generally have no specific expertise in the often complicated matter of scientific knowledge—are capable of assessing the proper use of science in the governmental decision-making.[14] One may wonder to what extent judges, if they are to move to a more intensive testing of how science is used for substantive policymaking, are in fact capable of understanding complex science. Specifically in the case of climate change, one can easily see that lawyers convened in courts, even if they are the highest qualified ones, usually have not been educated as climate scientists. This is the basis of our argument that judges should remain judges (Vos, 2013).

Role of judges in cases of fundamental rights and climate change

In the well-publicised Dutch *Urgenda Foundation v The Netherlands* case (see note 5), the court *ordered* the State of the Netherlands to achieve a more ambitious GHG reduction percentage than it is legally obliged according to EU law. The EU law dimension is a complex feature of this specific case and one which needs further clarification in future case law (Peeters, 2016). The Dutch Appeal Court took human rights (see **Chapter 11**), particularly *the right to life* and *the right to private life* as enshrined in the European Convention on Human Rights, as a legal basis, applying 'a particularly far-reaching interpretation of these standards' (Fahner, 2018).

This unprecedented judicial approach is truly unique. It clearly deviates from the judicial practice developed by the European Court of Human Rights which generally does not 'dictate precise measures which should be adopted by States in order to comply' with their human rights obligations (Preston, 2018: 158). Judicial adjudication in the case of climate change, particularly where it concerns the total emissions reduction to be pursued by a state, is problematic for another reason. It is hard to determine the direct relationship between GHGs emitted from one place and one time and the damage that may happen at another place at another time (see **Chapter 3**) (this also in view of adaptation measures that could have been implemented, for example against floods or heatwaves). In this vein, legal scholars have argued that it is hard to see how the extra reduction of emissions of a few percentage points by the Netherlands—as ordered in the Urgenda case—would significantly avoid the serious risk for the (Dutch) people of loss of life or disturbance of family life (Backes and van der Veen, 2018).

In sum, in view of the fundamental debate on how to strike a balance between democratic and judicial powers, it is questionable whether courts can legitimately award a claim for more stringent emissions reduction based on human rights provisions. By means of *prescribing* specific country-wide emissions reductions, such court decisions intervene into 'macro-environmental' decisions adopted by the democratically accountable legislator. On the other hand, human rights might have an important role to play for demanding specific governmental action to provide protection against natural disasters that find their roots in human-induced climate change, such as building dykes to be protected against floods.

Conclusion

Given the enormous and serious consequences that are projected to be caused by climate change, particularly if the increase of temperature rises (much) above 1.5°C, law is being called upon to play its role in the fight against climate change. Some voices are therefore calling for existing concepts such as causation in liability law to be adjusted to ease the possibility of letting polluters pay damages to victims of climatic disasters (Hinteregger, 2017). However, in our view this does not imply that courts should act as the primary standard setter on how to protect the climate system. The questions to be answered regarding the level and precise methods of emissions reduction are predominantly of a political nature, not a judicial one. Moreover, climate science and its embedded uncertainties are extremely complex for judges in courtrooms to understand.

Courts can hold states accountable for legally binding commitments, such as when an EU Member State does not implement a GHG emissions reduction target imposed by EU law. Courts are also important for judging the lawfulness of governmental penalties on emitters if they breach the emissions reduction obligations imposed on them. Victims can seek to use liability law to hold emitters to account for paying costs of the prevention and occurrence of damage related to human-induced climate change, although as yet it is an arduous task to achieve a successful outcome. Depending on claims submitted to them, courts can also force governments to collect and provide information on potential climate developments and possible risks. They can force governments to use adequate information and public participation procedures when developing their climate policies and legislation and decisions (Preston, 2018: 154; Aarhus Convention, 1998). Finally, courts can play a meaningful role in spurring governments to develop proper climate policies, particularly by requiring that those decisions are duly prepared and based on relevant information.

But there are important arguments against courts becoming the standard-setters for determining the level of emissions reduction and for deciding on the ways these reductions should be achieved. Courts have to test the legality of governmental decision-making, but they should refrain from becoming political decision-makers themselves.

Further reading

Bodansky, D. (2016) The legal character of the Paris Agreement. *Review of European, Comparative, and International Environmental Law.* **25**(2): 142–150.

This article assesses to what extent the Paris Agreement is legally binding on states—and therefore offers a useful perspective of the role of national courts in adjudicating on climate change policies. Bodansky argues that legal bindingness can be a double-edged sword if it leads states not to participate or to make less ambitious commitments. The Paris Agreement is significant beyond merely its legal status.

Fisher, E. (2013) Climate change litigation, obsession and expertise: reflecting on the scholarly response to *Massachusetts v EPA. Law and Policy.* **35**(3): 236–260.

This article reflects on why legal scholars have obsessed and become preoccupied with climate litigation in their research, focusing on writing about the high-profile US Supreme Court case of *Massachusetts v EPA.* Fisher probes the different narratives at play when legal scholars are reflecting on climate litigation and what legal expertise is implicated when they do so.

Peeters, M. (2016) Urgenda Foundation and 886 Individuals vs the State of the Netherlands: the dilemma of more ambitious greenhouse gas reduction action by EU Member States. *Review of European, Comparative & International Environmental Law.* **25**(1): 123–129.

This article offers a detailed assessment of the remarkable decision of a Dutch lower civil court to order the State of the Netherlands to reduce its greenhouse gas emissions by 25% by 2020. The court decision deals with the fundamental question of the extent to which a civil court can intervene in environmental decision making, particularly where this concerns the national policy of a European Union Member State.

Sabin Centre for Climate Change Law: Climate Change Litigation Database. Available at: http://climatecasechart.com. Accessed 7 July 2019.

This is an extensive database of court cases around the world involving climate change in some way. It divides the cases into public and private law claims and provides summaries of the cases and links to court documents where available. You might reflect on what definition of 'climate change case' is used in compiling this database.

Setzer, J. and Vanhala, L. (2019) Climate change litigation: a review of research on courts and litigants in climate governance. *WIREs Climate Change.* **10**(3): e580.

This is a wide-reaching overview, from a political science perspective, of academic research concerning climate litigation and courts. It makes arguments about where there is scope for future research concerning the conditions under which litigation informs climate governance.

Follow-up questions for use in student classes

1. What is the function of law and adjudication in relation to climate change policy? (Or why does law matter?)

2.   Are courts capable of deciding the emissions reduction level that a country should adhere to? Instead, what do you think of courts acting as 'informational catalysts' by assessing whether governments have properly made use of scientific information for their substantive decisions on the reduction of GHGs?

3.   Should the emissions reduction commitment of a specific country be made dependent on what other countries do? Suppose that certain countries do not adhere to the most ambitious GHG emissions reduction policies. Could that serve as a (legal) argument for a country with a comparable economic situation also not to adhere to an ambitious level of GHG reduction, but only to a comparable one?

4.   Who are most likely to be motivated to bring cases relating to climate change in the courts? (Think broadly!) Who is most likely to have funding to bring climate cases in the courts?

5.   What has been the most recent 'climate case' to capture global attention? Is this attention positive or negative, and is this justified?

Notes

1   *R (Spurrier) v Secretary of State for Transport and others* (2019) EWHC 2070. Available at: http://www.bailii.org/ew/cases/EWHC/Admin/2019/1070.html. Accessed 7 October 2019.

2   *Friends of the Irish Environment CLG v Fingal County Council* (2017) IEHC 695. Available at: https://elaw.org/friends-irish-environment-clg-v-fingal-county-council-2017-iehc-695-nov-21-2017. Accessed 19 July 2019.

3   *Plan B and [others] v Secretary of State for Business, Energy and Industrial Strategy* (2018) EWHC 1892 (Admin). Available at: www.bailii.org/ew/cases/EWHC/Admin/2018/1892.html. Accessed 19 July 2019.

4   *Ashgar Leghari v Federation of Pakistan* (WP No 25501/2015), Lahore High Court Green Bench – note orders of 4 Sept and 14 Sept 2015. Available at: https://elaw.org/PK_AshgarLe ghari_v_Pakistan_2015. Accessed 19 July 2019.

5   *Urgenda v The Netherlands* (2015) The Hague District Court. ECLI:NL:RBDHA:2015:7196. Available at: http://uitspraken.rechtspraak.nl/inziendocument?id=ECLI:NL:RBDHA:2015:7196. Confirmed on appeal in The Hague Court of Appeal, 9 October 2018. Available at: http://uitspra ken.rechtspraak.nl/inziendocument?id=ECLI:NL:GHDHA:2018:2610. Accessed 19 July 2019.

6   Ibid (Court of Appeal). At the time of writing (October 2019), Urgenda was heading for final appeal in the Supreme Court of the Netherlands, the highest court of appeal.

7   *Massachusetts v EPA* 549 US 497 (2007). Available at: https://supreme.justia.com/cases/fed eral/us/549/497/. Accessed 19 July 2019.

8   *Gloucester Resources v Minister for Planning* (2019) NSWLEC 7 (New South Wales Land and Environment Court). Available at: http://climatecasechart.com/non-us-case/gloucester-resources-limited-v-minister-for-planning/. Accessed 19 July 2019.

9   Moreover, according to the IPCC, limiting global warming to 1.5°C would require 'rapid, far-reaching and unprecedented changes in all aspects of society', and a temperature increase higher than 1.5°C exaggerates 'the risk associated with long-lasting or irreversible changes, such as the loss of some ecosystems'. IPCC Press release October 8, 2018. Available at: www.ipcc.ch/site/assets/uploads/2018/11/pr_181008_P48_spm_en.pdf. Accessed 19 July 2019.

10 For this essay we present our arguments in a simplified way and do not necessarily reflect the nuanced analysis we would develop in a more elaborated academic publication.

11 *Bettati v. Safety HiTech Srl* (1998) Court of Justice of the European Union. ECLI:EU:C:1998:353. Reports of cases 1998 I-04355.

12 *Société Arcelor Atlantique et Lorraine and Others v. Premier ministre, Ministre de l'Écologie et du Développement durable and Ministre de l'Économie, des Finances et de l'Industrie* (2008) Court of Justice of the European Union. ECLI:EU:C:2008:728. Reports of cases 2008 I-09895.

13 In some jurisdictions, courts act as regulators, such as in India and in Pakistan (Preston, 2018: 148–150). It falls beyond our expertise to discuss why this is legitimate in these specific countries.

14 For an example of the struggle judges face in understanding complex health or environmental issues see: Vos, E. (2004) Antibiotics, the precautionary principle and the court of first instance. *Maastricht Journal of European and Comparative Law.* **11**(2): 187–200. It merits further discussion of whether courts need to be assisted by appointed scientists and also further assessment of whether scientific advice has been duly taken in to account by the policymakers.

References

Aarhus Convention. (1998) *Convention on Access to Information, Public Participation in Decision-Making and Access to Justice in Environmental Matters.* Aarhus and Denmark: United Nations Economic Commission for Europe (UNECE).

Backes, Ch.W and van der Veen, G.A. (2018) Case note to the decision of the Court. *The Hague: Appeal State of the Netherlands v Urgenda*, 9 October 2018, nr. 200.178.245/01. Administratie-frechtelijke Beslissingen, AB 2018/417, issue 43.

Bähr, C.C.,Brunner, U., Casper, K. and Lusting, S.H. (2018) *Klimaseniorinnen*: Lessons from the Swiss senior women's case for future climate litigation. *Journal of Human Rights and the Environment.* **9**(2): 194–221.

Bergkamp, L. and Hanekamp, J.C. (2015) Climate change litigation against the states: the perils of court-made climate policies. *European Energy and Environmental Law Review.* **24**(5): 102–114.

Bouwer, K. (2018) The unsexy future of climate litigation. *Journal of Environmental Law.* **30**(3): 483–506.

De Sadeleer, N. (2016) The precautionary principle and climate change. In D.A. Farber and M. Peeters. eds. *Climate Law Encyclopaedia.* Cheltenham: Edward Elgar. Chapter 2, pp. 20–31.

European Union (2000) *Charter of Fundamental Rights of the European Union.* Brussels: Official Journal of the European Union, 18 July, C 364/01.

Fahner, J. (2018) Climate change before the Courts: Urgenda ruling redraws the boundary between law and politics [online]. Blog of the *European Journal of International Law.* Accessed 7 July 2019. Available at: https://ejiltalk.org/climate-change-before-the-courts-urgenda-ruling-redraws-the-boundary-between-law-and-politics/

Fisher, E., Scotford, E. and Barritt, E. (2017) The legally disruptive nature of climate change. *Modern Law Review.* **80**(2): 173–201.

Hinteregger, M. (2017) Civil liability and the challenge of climate change: a functional analysis. *Journal of European Tort Law.* **8**(2): 238–260.

IPCC (2018) Summary for Policymakers. In: *Global Warming of 1.5°C. An IPCC Special Report on the impacts of global warming of 1.5°C above pre-industrial levels and related global greenhouse gas emission pathways, in the context of strengthening the global response to the threat of climate change, sustainable development, and efforts to eradicate poverty* [Masson-Delmotte, V., P. Zhai, H.-O. Pörtner, D. Roberts, J. Skea, P.R. Shukla, A. Pirani, W. Moufouma-Okia, C. Péan, R. Pidcock, S. Connors, J.B.R. Matthews, Y. Chen, X. Zhou, M.I. Gomis, E. Lonnoy, T. Maycock, M. Tignor and T. Waterfield (eds.)]. World Meteorological Organization, Geneva, Switzerland, 32 pp.

Jasanoff, S. (2015) Serviceable truths: science for action in law and policy. *Texas Law Review.* **93**: 1723–1749.

Peel, J. and Osofsky, H. (2015) *Climate Change Litigation: Regulatory Pathways to Cleaner Energy.* Cambridge: Cambridge University Press.

Peeters, M. (2016) Urgenda foundation and 886 individuals vs The State of The Netherlands: the dilemma of more ambitious greenhouse gas reduction action by EU Member States. *Review of European, Comparative & International Environmental Law.* **25**(1): 123–129.

Petersen, N. (2017) The international court of justice and the judicial politics of identifying customary international law. *The European Journal of International Law.* **28**(2): 357–385.

Preston, B.J. (2018) The evolving role of environmental rights in climate change litigation. *Chinese Journal of Environmental Law.* **2**(2): 131–164.

Scott, J. and Sturm, S. (2007) Courts as catalysts: re-thinking the judicial role in new governance. *Columbia Journal of European Law.* **13**(3): 565–594.

Stone, C.D. (2010) *Should Trees Have Standing? Law, Morality and the Environment* (3rd edition). Oxford: Oxford University Press.

Türk, A.H. (2009) *Judicial Review in EU Law.* Cheltenham: Edward Elgar.

UNFCCC (2015) *The Paris Agreement.* Geneva. Available at: https://unfccc.int/sites/default/files/english_paris_agreement.pdf

Voigt, C. (2016) The potential roles of the ICJ in climate change-related claims. In D.A. Farber and M. Peeters, eds. *Climate Change Law.* Cheltenham: Edward Elgar. Chapter 13, pp. 152–166.

Vos, E. (2013) The European Court of Justice in the face of scientific uncertainty and complexity. In M. Dawson, B. de Witte and E. Muir, eds. *Judicial Activism at the European Court of Justice.* Cheltenham: Edward Elgar. pp. 142–166.

14 Does the 'Chinese model' of environmental governance demonstrate to the world how to govern the climate?

Tianbao Qin, Meng Zhang, Lei Liu and Pu Wang

Summary of the debate

This debate concerns the relevance and suitability of the 'Chinese model' of environmental governance for the rest of the world as other countries grapple with how to govern climate change. **Tianbao Qin** and **Meng Zhang** argue that heavily directed and enforced environmental regulation, as manifested in China, is an efficient and effective form of governance for bringing about the necessary structural and behavioural changes that can reduce greenhouse gas emissions. Other countries can learn from this model. **Lei Liu** and **Pu Wang** counter by arguing that China's record of environmental progress is oversold and, in any case, the 'Chinese model' is hard to transfer to other countries. Governing climate depends on the bottom-up interaction of multiple stakeholders at multiple levels, which is quite different from China's national practice of top-down governance.

YES: It offers a centralised governance model with a command-and-control approach (*Tianbao Qin and Meng Zhang*)

Introduction

There is a country that is an important developing country with the largest population in the world, 1.4 billion people. There is a country that is the second biggest economy in the world with its GDP accounting for 12.3% of world economic production. There is a country that has urgent need for economic growth to maintain its social stability, to cope with environmental challenges and to promote the recovery of the world economy from the shadow of the global economic recession; and whose GDP is growing at the rate of 7–8% per annum. Curious about which country can grab so many titles concerning global importance and economic prosperity? Yes, this country is obviously China.

However, these facts about China are not the entire story of the largest emerging economy. There is another crucial fact: it is also the largest greenhouse gas (GHG) emitter in the world, contributing about 27% of the total global CO_2 emissions (Qin and Zhang, 2016). Yet anyone who is interested in the global governance of climate change cannot neglect China's remarkable progress on environmental protection, its efficient environmental governance model and its unique experiences related to tackling climate change.

Climate change is a common challenge facing humankind and the Chinese government has always attached great importance to addressing it. It has promoted the construction of an equitable, rational, cooperative and win-win global climate governance system with a positive attitude. It has adopted practical and effective policies and measures to strengthen domestic actions to address climate change, demonstrating a firm determination to promote sustainable development and green and low carbon transformation. By taking a series of measures to push forward work on addressing climate change, China has become an important participant, contributor and torch-bearer in the global endeavour to seek an **ecological civilisation** (Fullerton, 2015).

Positive progress by China has been achieved. By 2017, China's CO_2 emissions per unit of GDP (i.e. its carbon intensity) had declined by approximately 46% compared to 2005. This already exceeded its 2020 target of reducing carbon intensity by 40–45% (Ministry of Ecology and Environment, 2018) and has reversed China's rapid growth trend in carbon emissions. Non-fossil energy accounted that year for about 14% of primary energy consumption, the task of afforestation and forest protection continues to advance and the ability of the nation to adapt to climate change has been continuously enhanced. The institutional mechanisms for addressing climate change continue to improve, organisation and team building have been strengthened and the awareness of climate change in the whole society improved (Ministry of Ecology and Environment (MEE), 2018).

These accomplishments in China's fight against climate change have been widely reported in recent decades, including its reduction of carbon emissions intensity, the fast development of new renewable energy and the recent launch of a nationwide carbon market (see **Chapter 6**). As the world faces the threat of climate change, people may ask whether the 'Chinese model' of environmental governance—which features top-down, **command-and-control** and campaign-style measures—demonstrates to the world how to govern the climate. As a responsible member of the international community, one which remains committed to providing solutions for global problems, China has actively responded to the climate challenge. It has vigorously promoted an ecological civilisation and achieved remarkable progress in pollution prevention and control to seek a future with bluer skies and clearer water.

Tragedy of the commons and command-and-control

It is said that climate change nowadays is 'the mother of all negative externalities'. GHG emissions have the attributes of public goods, which suggests the problem can be framed as a **tragedy of the commons** (Hardin, 1968). In accordance with Hardin's theory, command-and-control instruments are regarded as the most effective and efficient method to deal with the negative externality of environmental problems.

The tragedy of the commons syndrome, meshing economic growth and environmental protection, is often cited in connection with sustainable development, especially in the debate over climate change. It is a term used in social science to describe a situation of a shared resource system in which individual users acting independently according to their own self-interest behave contrary to the common good of all users by depleting or spoiling that resource through their collective action. The concept and phrase originated in an essay written in 1833 by the British economist William Forster Lloyd, who used a hypothetical example of the effects of unregulated grazing on common land (also known as 'a common') in Great Britain and Ireland. The concept became widely known over a century later due to an article written by the American ecologist and philosopher Garrett Hardin in 1968 (Hardin, 1968). In this modern context of environmental governance, the commons dilemma stands as a model for a great variety of resource problems in today's societies. These include the management of freshwater, forests, fish and non-renewable energy sources such as oil and coal, as well as global environmental issues such as climate change.

In economics, an externality is a cost or benefit that affects a party who did not choose to incur that cost or benefit (see **Chapter 5**). Negative externalities are a well-known feature of the tragedy of the commons. Economists often urge governments to adopt policies that 'internalise' an externality (Jaeger, 2012). In a typical example, governmental regulations can limit the amount of a common good that is available for use by any individual. Permit systems for extractive economic activities—such as mining, fishing, hunting, livestock raising and timber extraction—are an example of this approach. Similarly, limits on pollution are examples of governmental intervention on behalf of the commons. This is illustrated by some international legal instruments, such as the United Nations' Moon Treaty, Outer Space Treaty and Law of the Sea Treaty, as well as UNESCO's World Heritage Convention which involves the international law principle that designates some areas or resources the Common Heritage of Mankind.

To tackle the commons dilemma, command-and-control tools are often adopted as domestic environmental legislations and policies. Command-and-control (CAC) regulation can be defined as 'the direct regulation of an industry or activity by legislation that states what is permitted and what is illegal' (McManus, 2009: 32). The 'command' is the presentation of quality standards/targets by a government authority that must be complied with. The 'control' part signifies the negative sanctions (e.g. prosecution) that may result from non-compliance (Baldwin et al., 2011). In the case of environmental policy and regulation, the CAC approach strongly relies on the use of standards to ensure the improvements in the quality of the environment.

The CAC approach uses three main types of standards: ambient standards, emission standards and technology standards. Although these standards can be used individually, it is also possible to use them in combination. In fact, in most pollution control programs a combination of standards is implemented. The traditional model of command-and-control typically involves areas of environmental concern being dealt with by national governments. It is noteworthy that in China the most effective and efficient environmental governance approaches are introduced through the system of administrative law. The overwhelming majority of CAC regulation instruments are implemented through current environmental laws, a common strategy in Chinese legislation for the prevention and control of pollution (Qin, 2015).

Regulatory framework and implementation approaches in China

In China's current legal system, the environmental and energy policies and regulatory framework provide the legal basis for climate governance. China is a state where there is a long historical tradition of the ruling Communist Party operating in a central planning mode. Policies of the ruling party and central government therefore exert a great influence on national affairs, which can obviously be seen in the case of climate change.

In 2015, China formulated and submitted the 'Enhanced Actions on Climate Change: China's Nationally Determined Contributions' (NDC) to the UNFCCC, becoming the first developing country to submit such a document. It declared that China will achieve the peaking of CO_2 emissions around 2030, making best efforts to peak earlier, and that by 2030 it will lower carbon intensity (CO_2 emissions per unit of GDP) by between

60% and 65% from the 2005 level. This NDC provides a medium to long-term direction for China's work in combating climate change (Government of China, 2015). The 'Work Plan for the Control of Greenhouse Gas Emissions during the 13th Five-Year Period' was issued by the State Council in 2016 and is the most important policy guideline to address climate change in China during the 13th Five-Year Plan Period (2016–2020). This work plan will accelerate green and low carbon development, ensuring the fulfilment of low carbon development objectives and tasks set out in the Outline of the 13th Five-Year Plan.

In terms of a legal framework, China so far does not have any climate-specific legislation. Further legal developments will therefore be required in order to address many of the critical elements of a climate mitigation/adaptation enforcement regime. Perhaps more positively, in many instances China still has a strong foundation of existing environmental and energy law upon which climate-specific regimes may be established. The current domestic environmental and energy regulatory framework— such as planning approvals, pollution prevention and environmental impact assessment legislation—may all offer a basis for addressing climate change (Qin and Zhang, 2018).

During the last few decades, China has created a well-established framework of environmental legislation. The Constitution of the People's Republic of China acts as the foundation for this framework and the Environmental Protection Law of the People's Republic of China provides the main body of legislation. This incorporates two departmental branches of legislation: one to prevent and control pollution and the other to conserve nature and biodiversity. On the other hand, there are some new legislative progresses in addressing climate change. At national level these include:

- the 'Climate Change Law' and the 'Interim Regulations for the Administration of Carbon Emission Trading', both of which are being drafted;
- the 'Interim Measures for the Administration of the Clean Development Mechanism Projects';
- 'Measures for the Administration of the China Clean Development Mechanism Fund';
- 'Interim Measures for the Administration of Voluntary Greenhouse Gas Emission Reduction Transactions';
- 'Interim Measures for the Administration of Carbon Emission Permit Trading'.

These latter measures have all been steadily amended (Ministry of Ecology and Environment (MEE), 2018).

During the past decades, the Chinese Government has secured a series of positive achievements in strengthening the implementation of its Five-Year Plans and improving policy systems and mechanisms. The assessment of responsibility for GHG emissions control targets has been carried out at provincial level, enhancing laws, regulations and standards. All of these efforts are the best illustrations of the success of the Chinese environmental governance model with its top-down, command-and-control approach. The National Development and Reform Commission (NDRC) completed the 2016 examination and assessment of GHG emissions control targets of all mainland provinces (regions and municipalities). According to this assessment, 27 out of these 31 provinces had met their goal of yearly reductions in carbon intensity (Ministry of Ecology and Environment (MEE), 2018).

Under the 2018 institutional reform arrangements of the Chinese Government, the function of addressing climate change and emissions reduction was transferred to the newly founded Ministry of Ecology and Environment (MEE). This is an updated version of the former Ministry of Environmental Protection (MEP). In July 2018, the State Council adjusted the composition and personnel of the national leading group on climate change, energy conservation and emissions reduction in accordance with the new institutional setting, personnel changes and work requirements of the State Council. To implement the 'Work Plan for Greenhouse Gas Emission Control during the 13th FYP Period', local governments deployed the relevant work proactively. As of June 2018, all the 31 provinces (regions and municipalities) of mainland China had prepared their schemes or plans for GHG emissions control during the 13th FYP period. Among these 31 provinces, 25 of them had released their approved respective formal Work Plans for GHG emissions control during the 13th FYP period, while the proposed plans and schemes of the remaining six provinces were awaiting ratification by the provincial legislative bodies (Ministry of Ecology and Environment (MEE), 2018).

Analysis of the centralised command-and-control governance model in China

In general, the CAC approach to environmental governance has unique advantages that other types of legal measures cannot replicate when confronted with environmental problems such as climate change. The CAC approach adopts administrative commands and orders to directly control the environmental behaviour of economic entities and individuals, to formulate environmental standards that are legally compulsory and to plan ahead with clear and precise goals. This approach ensures rapid and efficient implementation, characteristics unmatched by other approaches to environmental protection, such as markets or voluntary instruments. The CAC approach is therefore characterised by certainty, authority, directness, efficiency and coerciveness for eliminating environmental externalities.

Second, for countries in social transition, the CAC approach plays a crucial and irreplaceable role in environmental governance. Emerging economies and developing countries, like China, are in a period of social transition. This transition brings about serious social disorders and imbalances, reflected in various worsening environmental problems like climate change. Dealing with these problems requires robust executive power to play an effective role within the legal framework. If this executive power were too easily challenged, it is likely that the administrative organs would be powerless to deliver good environmental governance in the transition period. The governance model of western countries—'big society and small government'—is therefore not suitable for countries like China which are in the process of major social transition. That is to say, countries undertaking painful socioeconomic reforms are in need of a powerful centralised government to effectively and lawfully implement CAC instruments to effectively and actively tackle challenging environmental problems such as climate change.

Third, highlighting the role and importance of the CAC approach does not mean that other types of environmental governance tools are not recognised, nor that flexible economic incentive mechanisms will necessarily be excluded from the package of environmental governance instruments. The variability, diversity and complexity of environmental problems calls for changes in the traditional model of environmental governance. Non-administrative mechanisms should be expanded to promote the diversification of governance methods. Under the government's good governance, central authorities *and* civil societies should together awaken environmental democratic self-determination and improve the market-oriented economic incentive mechanisms.

However, we must clearly understand that, due to the limitations of the market in the allocation of environmental resources, the market incentive mode of environmental governance has its weaknesses. The formulation of an economic incentive system is restricted by factors such as public acceptance, market feasibility and policy fairness (see **Chapter 6**). Any specific implementation of market mechanisms requires corresponding legal guarantees (see **Chapter 13**), especially the support of command-and-control instruments and regulations.

Last but not least, different countries are in need of diverse governance models in accordance with their own distinguishing conditions and environments. In order to effectively tackle climate change, technologies that support clean air, water and soil all over the world must gradually replace polluting alternatives. Implementation of this global transition will depend largely on local factors. What is the best approach for each government to take in promoting these technologies: the 'carrot' or the 'stick'? And which incentives or disincentives are most effective? Fixed and universal answers do not exist. But the Chinese experience of addressing climate change through the CAC approach offers a certain model for environmental governance. It also provides the world with a new perspective for how to discover and design a suitable and viable governance system for dealing with the various environmental problems arising from climate change.

Conclusion

Environmental policy, legal and standards systems for governing climate change have improved continuously in China in recent years. This has provided policy and institutional support for the improvement of climate change laws and regulations. The command-and-control approach has long been the major legal mechanism for environmental governance in China. The legal systems use methods such as a total control targets, discharge permissions and pollution control deadlines—and all three standards simultaneously. The CAC approach has played an important role for environmental governance and this has been helpful for climate improvement. In order to successfully implement climate targets embodied in the country's NDC, mandatory emissions-cutting measures enacted by central and local governments have proven effective to achieve low carbon goals. Moreover, with a CAC approach law enforcers have the authority to decide whether or not to inflict administrative punishment on polluters (Qin, 2015).

This Chinese model of environmental governance has taken root in China's unique political institutions, with a robust and dominant government and a hierarchical political system. It is right to question, as **Lei Liu** and **Pu Wang** do below, whether these Chinese experiences concerning environmental governance could be duplicated in their entirety in other countries. Nevertheless, given that climate change is a tragedy of the commons, we believe that a legally binding governance regime, based on the Chinese model and the lessons learned from it, is the most appropriate way to govern climate worldwide.

The Chinese Government has continuously played a positive and constructive role in international negotiations on climate change. It has taken a serious attitude towards its wider global responsibilities. China has strengthened multi-level consultations, engaged in international climate change dialogues with all countries and promoted consensus amongst relevant parties. The Chinese model of environmental governance plays an important role in promoting global climate governance and deepening international cooperation to address climate change.

NO: The Chinese experience is difficult to replicate by other countries (*Lei Liu and Pu Wang*)

Introduction

During the past four decades, China has been one of the fastest growing economies in the world. However, along with the rapid urbanisation, industrialisation and motorisation, China has to face daunting environmental and climate crises. Currently, China is the world's largest fossil energy consumer and carbon emitter and suffers from some of the most serious environmental pollution. In response to the increasing environmental pressure from both domestic and international actors, China has had to continuously enhance its environmental management system. In recent years, environmental law has been the fastest-growing type of legislation in China. In the transition to low carbon and sustainable development, China has secured a number of accomplishments. For example, according to the World Bank, China's carbon emissions per unit of GDP decreased by more than 70% between 1990 and 2014. At the same time, China has been the largest overall market for wind power since 2009 and accounted for a world-leading 35% of global cumulative wind installations in 2017 (Global Wind Energy Council, 2018). In 2017, China launched the world's largest carbon emissions trading market to facilitate the CO_2 emissions reduction of the country (see **Chapter 6**).

Global climate governance still faces serious challenges and whether and how the world will reach the 1.5–2°C target mandated in the **Paris Agreement** remains difficult to foresee. The world urgently needs innovative policy mechanisms to promote effective climate action, so people may wonder whether China's practice and achievement in environmental governance could be copied to the rest of the world. Before answering this question, it is necessary to introduce some background knowledge on China's political and bureaucratic system, which is the foundation of China's environmental governance.

It is impossible to understand China's environmental management system without understanding the country's political institutions. As shown in the simplified diagram of Figure 14.1, China's political system is a typical hierarchical structure with the central government on top. Below the central government are the 31 provincial-level governments of mainland China[1] and 26 national ministries which are in charge of specific public affairs, such as environmental protection, transportation, science and technology development. The provincial-level governments and national ministries report to central government and are accountable to it. Lower down the hierarchy are 2,851 counties and the lowest level includes 39,888 town governments. According to Chinese law, villages and street block organisations are not 'government' in legal terms, but are defined as self-governing grassroots organisations. However, in reality they always act as a level of government.

These lower-level government officials are appointed by higher level officials, rather than being locally elected, and their budgets largely depend on fiscal transfer from higher level governments. Thus, the lower-level governments are obliged to take orders from higher-level governments. Functional departments are not only administered by the upper-level government; they are also directed by the corresponding upper-level

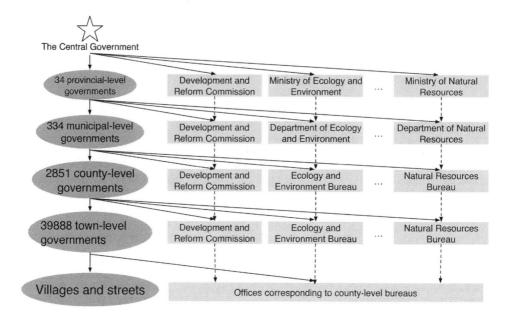

Figure 14.1 A simplified view of the political system of China

departments. Public policy is enforced through this layer-by-layer system, so that an idea in Beijing could in principle be transformed to action in each village of China.

Because of China's unique political environment and environmental management practice, this essay argues that the beneficial effects of a 'Chinese model', if any, are limited to the Chinese context and difficult to replicate for other countries.

China's environmental management accomplishments should not be exaggerated

The Chinese Government is attaching unprecedented importance to environmental protection and the low carbon transition and has achieved remarkable progress in recent years. But China's environmental progress may be oversold and it is too soon to conclude that China's environmental governance has been effective and successful.

First, China's progress in environmental governance has been reflected more in quantity indicators than in substantial quality improvement. Since the Ninth Five-Year Plan (1996–2000) for economic and social development, China has adopted the 'total amount control of pollutants' as the dominant environmental management strategy. This is implemented through a target responsibility system embedded in the top-down bureaucratic structure. That is, the higher-level government sets pollutant reduction targets for the lower-level government and links the performance in task fulfilment with certain punishments. This is usually the 'marking' of the political record of the lower-level cadres, which is supposed to influence the political career of these officials.

During past decades, according to official statistics, China has substantially cut down the amount of the prescribed important pollutants, such as sulphur dioxide and nitrogen oxides, but so far most Chinese cities are still suffering from severe air pollution. In 2016, only 84 out of China's 338 major cities (25%) had air quality that measured up to standard (Ministry of Environmental Protection, 2017). (It should be added that China's official air quality standards are poorer than those prescribed by the World Health Organization.) Severe air pollution has been estimated to cause 0.5–1.6 million premature deaths every year in China (Rohde and Muller, 2015). In addition to traditional haze pollution in winter, ozone pollution in summer is becoming a new salient issue. According to the Environmental Performance Index 2017 Report published by Yale University, the air quality in China ranked just 177th out of 180 countries.

Second, China's record of environmental improvement has not been widely recognised because of uncertain data quality. Although China has been claiming and trying to build a more transparent government and improving its environmental information disclosure, a number of studies have questioned the credibility of China's official statistics on carbon emissions and environmental quality from different level of governments. For example, Guan et al. (2012) reported an 18% difference in estimating China's carbon emissions when using two different official data sources. The 18% uncertainty equates to approximately 1.4 billion tons of CO_2—an amount greater than the total emissions of Japan (Guan et al., 2012). Ghanem and Zhang (2014) found that about 50% of cities officially reported dubious PM_{10} (inhalable particulate matters with diameters of 10 micrometres and smaller) pollution levels. Therefore, to some extent, the accomplishments of China in environmental management are still uncertain and unconvincing because inaccurate data may lead to wrong judgements.

Third, environmental protection should be regarded as an essential public service offered by the Government, but it has long been viewed as a burden for China's economic growth. The Chinese Government has argued for the slogan 'development is of overriding importance' in past decades and has pursued this aim consistently. Given the fact that China's per-capita GDP is still less than one-third of the USA level, China has been reluctant to sacrifice economic growth for environmental improvement and has attempted to arrive at a more balanced development pattern. In the course of this balancing, however, the environment has been sacrificed for the economy. Although China has largely reduced carbon intensity, its carbon emissions and energy consumption have rapidly placed the country at the top of the world league (Table 14.1). Being afraid that the reduction in absolute carbon emissions would cause an unacceptable economic cost, China has launched an intensity-based carbon market, which is more flexible in balancing emissions reduction and economic growth, but less committed to carbon abatement (Ellerman and Wing, 2003). In this respect, the EU is a more compelling governance role model: carbon emissions in the EU were reduced by 22% between 1990 and 2015, while the EU economy grew by 50% over the same period.

Costly bureaucracy and public policy processes

The 'Chinese model' of environmental governance has its roots in China's political institutions, which to a large extent are unique in today's world. Given the path dependency of institutional evolution, the 'Chinese model' is hard to transplant to other countries, which usually do not have a centralised regime covering such a huge population and land area.

First, institutional failure tends to happen when policy making and implementation are completely organised in a top-down manner. This is especially so when operating at

Table 14.1 The top ten CO_2 emitting countries and regions (2016)

Country/region	CO_2 emissions (billion tons)	Country/region	CO_2 emissions per capita (tons)
China	9.10	Qatar	30.8
United States	4.83	Curaçao	25.9
EU (28)	3.19	Kuwait	22.2
India	2.08	Bahrain	20.8
Russia	1.44	United Arab Emirates	20.7
Japan	1.15	Gibraltar	19.0
Korea	0.59	Saudi Arabia	16.3
Iran	0.56	Australia	16.0
Canada	0.54	Trinidad and Tobago	15.5
Saudi Arabia	0.53	United States	15.0

Source: IEA CO_2 Emissions from Fuel Combustion, OECD/IEA, Paris, 2018.

a large-scale with five levels of government and diverse socioeconomic conditions across different regions. When the policy arrangement from the higher-level government does not fit with local conditions and local government does not have veto power to resist the arrangement or even to negotiate, the efficacy of the policy tends to be discounted or even to contravene the original policy objective. In a multi-level polity, the incentive incompatibility between central and local governments is a common problem, particularly when the power distance between different layers is relatively large like China. Meanwhile, policy information is very easily distorted when going through the layers. The more layers the information passes down, the bigger is the misunderstanding and misinterpretation. It is conceivable and imaginable that the central government may not even be completely aware of what is happening at the grassroots level when making rules for them. Local government also tends to complain that the higher-level policy is too arbitrary and allows for no consideration of regional differences and local conditions.

For central government to avoid the distortion and local manipulation in information delivery, it has to invest huge administrative resources to supervise local government. For local governments, the frequent supervisions from the upper-level government always costs them a lot of time in preparation (or manipulation) of materials to prove their merit or working efficiency. This interrupts their regular work. Moreover, this supervision *per se* has a large risk of being ineffective. Local government has an absolute information advantage about local realities, which constitutes the most important bargaining power when gaming with upper-level government. In addition to regular supervision, the central government needs to perform stringent and large-scale special supervisions to overcome the information asymmetry.

When facing such supervision, in order to improve the environmental quality in the shortest time, many local governments take extreme measures, such as forcibly closing all the factories or cutting power supply. This can be seen with the interim measures taken before China's several mega-events, such as the 2008 Olympic Games, the 2014 Asia-Pacific Economic Conference in Beijing and the 2016 G20 Conference in Chengdu and Hangzhou. The campaign-style 'environmental storm' generated before such events is effective for hitting the target of improved environmental quality over the short-term. But the radical measures relying on administrative power often ignite public discontent, especially at the local-level. People suddenly lose their jobs and property without them violating any environmental law and without enough sense of patriotism to be persuaded that their sacrifice is necessary.

Second, China's public management adopts the principle of 'territorial management', which can simply be understood as allocating responsibility according to the administrative territory. In a top-down and multi-layer system, territorial management easily becomes an exercise in buck-passing where the lowest level has to bear the biggest pressure. A commonly recognised viewpoint on China's bureaucracy is that 'the grassroots government cannot retain talent'. The large workload and high mental pressure grassroots environmental officials experience, combined with poor salaries, seriously affects their working motivation. During the nationwide 'environmental storm' commanded by the central government in 2018, there was even a local environmental official who died from overwork. Furthermore, in this hierarchical system, local cadres pay more attention to their superior officials than to the citizens in their administration. Evaluated against the public interest, they usually care more about their political career—which is decided by higher officials—than they do about the public. On many occasions, the primary goal of local policy implementation is therefore to please and game the higher-level government instead of improving public welfare.

The third reason China's environmental governance system is costly and inefficient is the problem of enforcement. The country's environmental legislation system has experienced a great leap forward during last decades and become comprehensive and complex. But its enforcement often falls short of a legal standard. Local environmental agency is subject to the control of local government on budgets, manning quotas, as well as on the appointment and demotion of the agency heads. So, when environmental law enforcement bears down on large projects and enterprises in a local jurisdiction, it is always interfered with by local government and other economic and development agencies. This tends to require loosening of environmental standards for the sake of economic interest. Similarly, at national level, the protectionism offered to central state-owned enterprises results in connivance with their environmental violations, resulting in a large number of serious pollution incidents.

Non-applicable lessons on the global scale

The top-down model of environmental governance might work for a single country such as China, but at the global level it is ineffective because there is no powerful and superior third party to enforce the climate action of sovereign nations.

As a global public good, managing climate change cannot be addressed without global cooperation and action. This depends on each country's rational decision-making based on the national interest and on an equitable distribution of responsibility between countries (see **Chapter 10**). So far, all the climate actions taken by countries have been voluntary. If any country is unwilling to curb its carbon emissions or even wants to quit a signed international climate agreement, as the USA seeks to quit the Paris Agreement, there will be no substantial punishment other than verbal criticism of that nation's so-called international image. Therefore, *global* climate governance can only be conducted via negotiation and interaction between stakeholders, i.e. a bottom-up approach.

A once-for-all global emissions reduction agreement implemented by national governmental units is theoretically plausible, but has been proved to be very difficult to implement because of the difficulty of compromising the interest conflicts between countries.

Ostrom (2009), however, argued for the important role of smaller-scale agency and put forward a polycentric solution to global climate change. By this she meant actions taken at various levels with active oversight by local, regional and national stakeholders. In the Chinese context, however, the state remains the dominant actor in environmental governance. **Non-state actors**, including the market and civil society, have been peripheral, although both of these actors have been argued, and proven, to be very important in public management (see **Chapter 12**).

For *market actors*, the distinction between market-based and regulatory policies for environmental management is blurred. The state should act as the role of regulator and enabler of the policies, but in many cases it excessively intervenes in market activities (Wang et al., 2018). In regards to *civil society actors* such as NGOs, they remain primarily a means for policymakers to reach their own goals, acquire ideas and soften opposition from the public sphere (Moe, 2013). The fundamental strategy of the Chinese Government with respect to NGOs is to encourage their professional capacity, but to discourage political mobilisation capacity. Although seemingly promoted through a series of decrees in China, such as the 'Provisional Measures on Public Participation in Environmental Impact Assessments', the engagement of the general public and social organisations in local environmental management is fairly limited. This has led to numerous public protests. People neither acknowledge the relevant information nor have the legal rights to refuse the construction of polluting projects in their neighbourhood.

In a nutshell, the Chinese Government possesses significant social and political resources and has mandatory power to force multiple stakeholders to take action according to the intention of key decision-makers. However, the global context is completely different. Here what is needed is either diplomatic negotiation between national actors, or else decentralised bottom-up actions taken by diversified actors at multiple scales and multiple levels. Contrary to the suggestion of **Tianbao Qin** and **Meng Zhang** above, the 'Chinese model' of environmental governance is not applicable if considered from this global perspective.

Conclusion

China can demonstrate to the world that it has progressed on environmental governance, but so far the 'Chinese model' is difficult to generalise for the practices of other countries or to scale-up to the global level. First, China's environmental governance is far from a role model. The current progress appears successful in some figures, but the country still cannot meet the requirement of 'measurable, reportable and verifiable' environmental improvement. The actual improvement in environmental quality and carbon emissions control is also limited. In the face of economic growth, environment has always been a secondary concern, although the current authority is trying to change such traditional ideas and behaviour.

Second, the constitutional and institutional context underlying China's environmental governance is hard to transplant to other countries, both politically and economically. A hierarchical system with large power distance between different governance levels tends to distort policy information going through the system. This leads to the deviation of policy objectives and failures in rational policy-making. The prevention of such distortion relies on top-down supervision, which is very costly and has limited effect. In addition, in a strict hierarchy the lowest level of government is the weakest, but the most stressful to work in. Each cadre of officials pays more attention to their superior officers than to the citizens in their administration. The primary goal of local policy implementation is to please and game the higher-level government instead of improving public welfare.

Last, but not least, and different from the Chinese context, fighting climate change at the global level requires the collective action of all countries, which depends on equitable interaction between actors and self-governance. All sovereign nations are independent and there is no 'global central government' to unilaterally enforce climate policy. In the perspective of Ostrom's polycentric model, China's political and legal environment does not favour decentralised bottom-up action and participation of multiple stakeholders at multiple levels. The active and effective participation of multiple stakeholders—*inter alia*, non-state stakeholders including NGOs, media and the general public—is still limited and marginalised due to the prevailing political sensitivity. The role and decision-making of all the stakeholders are largely controlled by central government policy.

To sum up, China needs to distil more generic innovations in environmental governance to make a contribution to collective global climate action of the world. Nationally, its ongoing environmental reforms, particularly its goals for a low carbon economy, have been regarded as a viable national strategy offering new impetus for the economic development of the country. For other countries, and for the entire world, there is no simple solution to mitigating climate change. Every country should keep in mind the threat of climate change and associated environmental deterioration, and search for suitable governance solutions congruent with their local physical, political and social conditions.

Further reading

Fiorino, D. J. (2018) *Can Democracy Handle Climate Change?* Cambridge: Polity Press.

Global climate change poses an unprecedented challenge for governments across the world. In this short book, Fiorino examines the assumptions and evidence offered by sceptics of democracy and its capacity to handle climate change, in contrast to more authoritarian regimes. He concludes that democracies typically enjoy higher levels of environmental performance and more effective climate governance than do autocracies.

Liu, J. C-E., Huang, H. and Ma, J. (2019) Understanding China's environmental challenges: lessons from documentaries. *Journal of Environmental Studies and Sciences.* **9**(2): 151–158.

Films can be powerful tools in teaching environmental studies and this article examined 36 environmental documentaries on China for their content, theme, production and reception with a view towards using them as teaching tools. It can serve as a 'viewing guide' for students and instructors to learn about China's environmental challenges through films.

Liu, L., Wang, P. and Wu, T. (2017) The role of nongovernmental organizations in China's climate change governance. *WIREs Climate Change.* **8**(6): e483. doi: 10.1002/wcc.483.

The role of NGOs in domestic climate change governance differs across countries due to varying political, legislative and cultural contexts. China is engaged in the challenging process of low carbon development, which may not be achievable through exclusive reliance on top-down management and voluntary actions by the private sector. This article reviews the participation of NGOs as a civil society actor in China.

Lo, K. (2015) How authoritarian is the environmental governance of China? *Environmental Science & Policy.* **54**: 152–159.

This article challenges the prevailing perception that the environmental governance of China is a case exemplar of authoritarian environmentalism. Using low carbon climate governance as an example, it shows that although China's national low carbon policy appears highly authoritarian, the situation on the ground is much more ambiguous, displaying a mixture of authoritarian and liberal features.

Qi, Y. and Wu, T. (2013) The politics of climate change in China. *WIREs Climate Change.* **4**(4): 301–313.

This article explains the evolution of climate policy-making and implementation processes and mechanisms in China over the past two decades. It shows how the policy-making process is characterised by consensus building at the centre, in contrast to the open public and partisan debates in Western democracies.

Qin, T. and Zhang, M. (2017) Development of China's environmental legislation. In E. Sternfeld, ed. *Routledge Handbook of Environmental Policy in China.* Abingdon: Routledge. pp. 17–30.

In this book chapter, the authors show how China's attitude to environmental issues is derived from its own distinctive cultural and historical features. They argue that China's environmental legislation should be seen as a social experiment, focused on addressing environmental problems by triggering broad societal reform that will affect people's everyday lives by modifying their ways of thinking and behaviour.

Follow-up questions for use in student classes

1. Compare China's political system for public policy-making and implementation with that of your own country. What lessons can your country learn from the Chinese experience of regulating greenhouse gas emissions?
2. What do you think are the advantages and disadvantages of China's centralised bureaucratic governance system for managing climate change?
3. How would you facilitate and promote public participation in China for fighting climate change?
4. What role should China play in addressing climate change in the era of the Paris Agreement on Climate Change?

Note

1 In addition are the three non-mainland provinces of Hong Kong, Macau and Taiwan.

References

Baldwin, R., Cave, M. and Lodge, M. (2011) *Understanding Regulation: Theory, Strategy and Practice.* 2nd edition. Oxford: Oxford University Press.

Ellerman, A.D. and Wing, I.S. (2003) Absolute versus intensity-based emission caps. *Climate Policy.* **3**: S7–S20.

Fullerton, J. (2015) China: ecological civilization rising? [online]. *HuffPost,* 2 May. Accessed 2 June 2019. Available at: www.huffpost.com/entry/china-ecological-civiliza_b_6786892?guccounter=1

Ghanem, D. and Zhang, J. (2014) 'Effortless perfection': do Chinese cities manipulate air pollution data? *Journal of Environmental Economics and Management.* **68**: 203–225.

Global Wind Energy Council. (2018) *Global Wind Report: Annual Market Update 2017.* Brussels: GWEC.

Government of China. (2015) *Enhanced Actions on Climate Change: China's Intended Nationally Determined Contributions*. Beijing: National Development and Reform Commission of China. Available at: https://www4.unfccc.int/sites/ndcstaging/PublishedDocuments/China%20First/China %27s%20First%20NDC%20Submission.pdf.

Guan, D., Zhu, L., Geng, Y., Lindner, S. and Hubacek, K. (2012) The gigatonne gap in China's carbon dioxide inventories. *Nature Climate Change*. **2**(9): 672–675.

Hardin, G. (1968) The tragedy of the commons. *Science*. **162**(3859): 1243–1248.

Jaeger, W. (2012) *Environmental Economics for Tree Huggers and Other Skeptics*. Washington, DC: Island Press.

McManus, P. (2009) *Environmental Regulation*. Sydney: Elsevier Ltd.

Ministry of Ecology and Environment (MEE). (2018) *China's Policies and Actions for Addressing Climate Change*. Beijing: Ministry of Ecology and Environment.

Ministry of Environmental Protection (MEP). (2017) *China Environmental Condition Bulletin 2016*. Beijing [in Chinese]: Ministry of Ecology and Environment.

Moe, I. (2013) *Setting the Agenda. Chinese NGOs: Scope for Action on Climate Change*. Lysaker. Norway: Fridtjof Nansen Institute.

Ostrom, E. (2009) *A Polycentric Approach for Coping with Climate Change Policy Research Working Paper 5095*. Washington: World Bank.

Qin, T. (ed.) (2015) *Research Handbook on Chinese Environmental Law*. Cheltenham: Edward Elgar Publishing.

Qin, T. and Zhang, M. (2016) Greenhouse gas emissions trading in China. In D. Farber and M. Peeters, eds. *Encyclopedia of Environmental Law: Climate Law*. Cheltenham: Edward Elgar Publishing. pp. 400–414.

Qin, T. and Zhang, M. (2018) *Climate Change and the Individual: A Perspective of China*. National Report for International Academy of Comparative Law Conference in Fukuoka, Japan.

Rohde, R.A. and Muller, R.A. (2015) Air pollution in China: mapping of concentrations and sources. *PLOS ONE*. **10**(8): e0135749.

Wang, P., Liu, L. and Wu, T. (2018) A review of China's climate governance: state, market and civil society. *Climate Policy*. **18**(5): 664–679.

15 Are social media making constructive climate policymaking harder?

Mike S. Schäfer and Peter North

Summary of the debate

This debate revolves around digital communications and social movement theory and introduces students to arguments about the relationship between social media, democracy and climate change. **Mike S. Schäfer** argues that the fragmentation of established news media brought about by social media and the associated prevalence of online echochambers fuels identity politics and issue polarisation. This makes climate policymaking considerably harder and undermines the possibility of a constructive deliberative democracy. **Peter North** disputes this argument, pointing out the new possibilities afforded by digital communication platforms for coalescing new social movements which can speed the transmission of counter-cultural values which may be necessary for climate mitigation. Social media are good for climate democracy.

YES: Social media fragment, polarise and worsen climate change communication (*Mike S. Schäfer*)

Introduction

The rise of online and social media has altered public communication significantly. More people, and more diverse people, have access to the internet around the world and in different strata of the respective populations—among young and old, men and women and the well and the less well educated. In the process, online and social media have become one of the most important places where people inform themselves about issues they consider important. They consume news to a substantial degree via these media nowadays, particularly, and increasingly, via social media such as Facebook, YouTube, Instagram, Twitter and Reddit. This is also true for climate change and climate politics, issues which are complex and detached from many people's lives and which reach many of them via media. Among these media, social media have risen strongly in importance in recent years.

On the one hand, this emergence and rise of the internet and particularly of social media has implications for individuals. Content delivered via social media platforms shapes the agenda of issues many people concern themselves with, influences what they perceive as important, how they frame and subsequently approach issues and what sources they consider as trustworthy in informing and helping them. On the other hand, social media are also of larger, sociopolitical importance. In western democratic countries they have become important intermediaries in the political process. In these countries, politics has to answer to the public—to legitimate its policies in public—and to search for approval in public opinion, at least in the long run. Social media have become an important arena where these processes take place. In more authoritarian countries, social media also play an important, albeit different role. In such countries they can serve to catalyse counter-publics which draw attention to underrepresented and critical issues and allow for a considerable degree of open deliberation, as has been shown for environmental issues in China (see **Chapter 14**).

There are positive and negative perceptions of these implications. *Cyber-optimists* have viewed them largely positively. They have emphasised, for example, that online and social media could improve the communication of science to the broad public and do so in a way that enhances user understanding and engagement. In contrast, *cyber-pessimists* have stressed the dangers imminent in social media. They have emphasised that social media are mostly used for entertainment and 'soft news' instead of more serious topics and that they cater to individual interests instead of common concern. They also point out that many people have difficulties distinguishing reliable from unreliable information on social media platforms and that social media might cause a fragmentation or even polarisation of public debate.

Unfortunately, the growing body of social-scientific research on online communication about climate change, and on the political options to mitigate it, suggests that the pessimists' view is more apt. The rise of social media is making climate policymaking considerably more difficult. And this seems to be more than the usual cultural pessimism and critique that has previously surrounded the development of new media. Yes, the advent of new media technologies in the past has often been greeted with hesitation, concern or outright rejection—from popular novels to broadcasting media to the home computer. And yes, scepticism about climate change and the necessity of impactful climate change policymaking has always been present to some extent, particularly in anglophone countries. But social media are disruptive in ways and to a degree that differ from historical examples. They are 'a transformative digital technology' that has broken 'down many of the barriers to individuals communicating with each other' and 'disrupted established hierarchies of communication … as never before' (Pearce et al., 2019: 1). This is cause for concern.

How social media undermine public communication about climate change and climate politics

Social media drive climate change communication from pluralisation to fragmentation to polarisation

Structurally, social media are conducive to a pluralisation of communication. Opening a Facebook, Instagram, Snapchat, YouTube or Twitter account is easy. Once users are there, they can easily post content themselves or like and share the posts of others. Originally, this gave rise to hopes that social media could empower previously underprivileged individuals and groups and give underrepresented views and arguments more room in public debate.

At first, there seemed to be something to these hopes. Social media indeed seemed to pluralise public debates, including those about climate change. They were described optimistically as allowing previously less visible groups and arguments to become more prominent in public debates (for a summary of the literature see Schäfer, 2012). But more recent research has tempered this optimism, showing that social media do not empower all groups equally. On the one hand, scholarship 'provides a consistent picture of mainstream media sources remaining prevalent' (Pearce et al., 2019: 5). But on the other hand, among the remaining non-mainstream voices, climate sceptics denying the existence, anthropogenic origins or impacts of climate change seem to profit most and are particularly prevalent in social media (for an overview see Pearce et al., 2019).

This communicative bifurcation seems to result in a fragmentation, and deepening polarisation, of climate change communication in social media. It has led to the emergence of **echo chambers**—where users construct their online communication in a way that lets them encounter positions which are mostly in line with their pre-existing positions—and to **filter bubbles**—where pre-existing user preferences are reinforced by the algorithmic selection and presentation of content that is in line with these preferences. Even though research indicates that talk about filter bubbles and echo chambers may be somewhat exaggerated and may not apply to all political issues, studies on climate change communication paint a different picture. One study from 2015

> used network analysis to investigate segregation and interaction between communities of Twitter users, categorizing users as activist, skeptic, neutral, unknown, ambiguous or unclassified according to tweet content. They found high levels of polarization and a tendency for active users (either skeptic or activist) to have strong attitudes, leading them to conclude that Twitter climate change discussions are 'characterised by strong attitude-based homophily and widespread segregation of users'.
>
> (Pearce et al., 2019: 6)

Other studies paint a similar picture.

This is problematic because it makes public consensus about the existence of climate change and about the best direction for policymaking more difficult. It also may give many people a wrongful sense of empowerment. They may perceive *some* variety of voices and standpoints in their social media feeds on climate change and have the impression they are getting a comprehensive overview of the perspectives that exist in society. Yet they might not actually encounter truly opposing views.

Social media erode the cognitive basis of climate politics

A second point of concern is the status of (climate) science in social media. Science communication research has demonstrated that the simple transmission of scientific findings will neither reach a large share of the public nor change its attitudes or behaviours towards issues like climate change (see **Chapter 9**). But the status of science in social media communication about climate change is worrying for three reasons.

First, social media communication is generally focused more on entertaining 'soft news' topics than on 'serious' news. It is difficult to assess the overall portfolio of social media, even for single platforms such as Facebook or Twitter. Yet it is disconcerting that some research shows that the most prominent Facebook topics, YouTube movies or Instagram posts are concerned with neither politics or science. This is because these platforms have become important sources of information and orientation for many people and have achieved this status at the expense of legacy news media.

For some people this is not problematic. Those segments of the population who *are* interested in climate change and avidly search for information about the issue will find an abundance of information about climate change and climate politics on social media. This is more so, and more comprehensively so, than ten or 20 years ago. But it is different for users with less or little interest in climate change or environmental issues and who do not actively search them out. Those users—who make up the majority of the population in many countries (see the overview on audience segmentation analyses by Hine et al., 2014)—might have encountered these issues ten or 20 years ago while routinely reading their morning newspaper or watching the evening news. But they are less likely to encounter them nowadays via social media.

Second, scientists themselves are not strongly present on social media. Many of them use social media for professional purposes or in their personal lives and some prominent examples of climate scientists participating in social media debates do exist (particularly in blogs and on Twitter). But overall, scientists are not strongly present in social media.

Third, the representation of climate science in social media varies strongly. Scholars have repeatedly analysed the representation of climate science in social media and shown that these representations are rather diverse. While many accurate accounts of the science underlying climate change can be found, for example around the publication of the Assessment Reports of the IPCC, (climate) science is also often misrepresented by climate sceptics or corporate PR (for a summary see Schäfer, 2012: 533f). These misrepresentations profit from the prevalent algorithmic curation on social media which is designed to maximise user interaction in order to better sell advertisements. It is *not* designed to maintain or improve information quality or accuracy.

Social media lower the communicative standards of the public sphere and make compromise more difficult

Scholars of public debates have often used public sphere theory to assess the quality of societal communication, often following a **Habermasian ideal** (see Edgar, 2006: 64–67). This ideal advocates for the inclusion of diverse actors in all public debates. As we have seen, this seems only partly realised in climate change communication on social media. They are also concerned with the style of communication, arguing for a rational, civil and respectful debate. Only such conditions, these scholars argue, can give rise to fruitful deliberation and, subsequently, to a robust consensus about how to tackle collectively relevant problems such as climate change. To the contrary, however, many scholars have worried that social media—in which gatekeepers such as journalists who present content based on professional norms and established procedures are less important or even absent—would give rise to more incivil debates, less dialogue, bullying and *ad hominem* attacks. After all, users can participate in social media discussions while remaining anonymous, using them to advance their own agendas.

Research on social media communication surrounding climate change indeed suggests that the lofty demands of public sphere theory are not met. While this research demonstrates that viewpoints and arguments in these discussions are indeed diverse—a fact that would be evaluated positively by public sphere theorists—it also shows that they leave much to be desired otherwise (for an overview see Schäfer, 2012). Social media discussions about climate change have been shown to be unstructured, polarised and often ideologically driven:

They are often long[,] unstructured, angry or abusive, and filled with assertions that would be difficult to cross-check[. T]here are high numbers of controversial and uncheckable assertions, plus more than a few questions with no obvious answers, or answers with no obvious questions. Entries are often highly disjointed and difficult to follow—part polemic, part rant, part ramble, part squabble, and often involving people flatly contradicting or sniping at one another. The calibre and tone of content is often 'uninspiring', and can in places descend to playground level.

(Gavin, 2009: 137)

Research has also found that incivility and sarcasm are frequently used in social media communication about climate change, particularly from climate sceptics.

Social media undermine the economic basis of established communication intermediaries, such as news journalism

The rise of social media also has indirect detrimental effects on other societal intermediaries who are concerned with providing information and orientation on issues of collective relevance. This is primarily true for news journalism. Legacy media have come under pressure in many countries (for overviews see Schäfer, 2017), with declining audiences and subscriptions and, subsequently, advertising revenue shrinking. This has led many publishers to reduce costs and these cuts hit specialised science and environmental journalism hardest. After all, science and environmental journalism never was a topical priority for many media. When facing economic difficulty, 'it is all too often the case that science news is regarded as expendable', i.e. 'a luxury increasingly difficult to justify when certain other types of news will be both cheaper to produce and more popular with audiences (and thus advertisers)' (Allan, 2011: 773). As a result, many journalism desks with topical expertise on climate change were reduced in size or even closed. In addition, working conditions for the remaining journalists are getting worse. They have to deal with an increasing amount of information, including social media content, and have to do so with fewer staff, fewer resources and shortened response times. Yet they still have to produce more content for a growing number of channels, ranging from their journalistic core products to online and mobile publications to social media (for an overview see Schäfer, 2017).

For a few years, online-born journalistic players such as Buzzfeed and Vice, who were strongly present on social media and aiming to use their communicative logic, were seen as a rising alternative to established news journalism. And their coverage had indeed been shown to produce a different, potentially more impactful climate change coverage that could reach new audiences in engaging ways (Painter et al., 2016). But many of these players have also run into serious economic difficulties and hopes have subsided. The result is worrying. People face a quickly changing media landscape and an ever-growing information environment where they have to navigate content that is diverse in positions and quality. This content is accompanied by new contextual cues such as likes, shares, comments and so on that influence credibility judgments. Social media have helped erode the institutions that have historically provided orientation.

As a result, the basis for comprehensive climate change policymaking looks bleak. Social media will give fewer people the impression that climate change is a real and pressing issue; they provide them with one-sided views and they give the impression that other perspectives do not exist or are less viable. They may even make them more used to an aggressive style of communication with people they disagree with. Under such conditions, having constructive discussion and finding societal consensus—important preconditions for successful and sustainable policymaking—is getting exceedingly difficult.

Conclusion

Many of the claims made in this essay have to be cautioned in two ways. First, social media are a quickly changing object. With many different platforms emerging every year, with audiences navigating and shifting quickly between these platforms, and with platforms themselves often changing or adopting new features, social media present a moving target for analysis. Second, there are still not enough analyses of social media communication about climate change and its links to policymaking and those studies that do exist have serious biases—like focusing mostly on anglophone countries and predominantly on Twitter (cf. Pearce et al., 2019). Accordingly, the empirical basis for many of the claims above should be tested more widely.

Still, the existing scholarship paints a picture that is worrying. Social media communication about climate change is often polarised and incivil. It may give people an incomplete picture of the issue at hand and isolate or even alienate them from groups with different views. In addition, it has detrimental effects beyond social media themselves. It erodes environmental and science journalism which is in a substantial crisis. It will take a long time before new and sustainable models of science journalism will develop.

These developments thoroughly undermine the social and discursive basis of climate change-related policymaking. Particularly in democratic countries, policymaking on any issue has to rest on public debate and societal legitimation. Current trends in social media make it less likely that such debates on climate change will occur. They are also likely to decrease the acceptance for policies that may advance measures not in line with users' own viewpoints and make social consensus-building, which should be the foundation of any successful policymaking, more difficult.

NO: They allow new, radical voices into the debate, thereby facilitating democratic forms of policymaking (*Peter North*)

Introduction

Concerns about the so-called abuse of information and communication technologies (ICTs, including the Internet, smartphones, texting, Twitter, Instagram, social media and social networks) are prevalent in early twenty-first century life. This is especially so in wired, democratic societies that put a high value on freedom of speech in a liberal marketplace of ideas. Much of this anxiety reflects concerns we see regularly with the introduction of any disruptive new technologies because the losers—in this case the mainstream, paid-for media industries, policymakers used to making policy in closed, controlled environments—have to cope with new entrants. Some of these concerns are implicated in scares about the use of social media for nefarious purposes. And some of them result from confusion between the right of everyone to their opinion—which is free—and the right to construct their own facts, which is not.

When attempting to respond to a complex grand challenge like avoiding dangerous climate change, or adapting to inevitable climate change already in the system, there are concerns that these new social media make constructive climate policymaking harder. This can be a result of anxiety about the proliferation of 'bad' or 'inaccurate' science online produced by climate sceptics, perhaps based in neoliberal think tanks supported by fossil fuel interests (see **Chapter 9**). These concerns notwithstanding, in this essay I dispute the argument put forward earlier by **Mike S. Schäfer**, that social media make constructive climate policymaking harder. I do so for four main reasons: (1) contrarian opinion is not new; (2) scientific knowledge is not rolled out unchallenged from the lab; (3) what to do about climate change is value-laden, not fact-driven; and (4) both citizen scientists and social movements can constructively add to political and policy debates using social media.

Contrarian opinion is not new

First, it is wrong to argue that social media have completely changed the relationship between official, elite knowledges about how the world works and subaltern challenges to it. For as long as there have been social problems, people have advanced ideas about what is wrong with the world, what needs to change and how this should happen. They have organised to effect this change using a number of what **social movement theory** scholars call 'protest technologies' through which oppositional ideas are developed and dispersed through society. These can include grassroots or 'samizdat' newspapers, pamphlets and leaflets produced on clandestine and independent presses; letters to newspapers or between individuals or groups of likeminded individuals; independent journalism; proclamations (perhaps the primary example being Martin Luther's 95 theses nailed to the church in Wittenberg in 1517 and thereby sparking the Reformation); speeches (on the street, at a rally, in Parliament or a local authority); and sermons (in religious buildings). ICTs are merely the latest development in communication technology, as revolutionary as the printing press, enabling new ideas to spread. But they are not fundamentally new, even if the process is now quicker and more ubiquitous.

Sometimes new ideas are rejected by elites. Galileo was placed under house arrest for suggesting that planet Earth orbited the sun and Darwin was ridiculed by some for proposing evolution through natural selection. Sometimes these new ideas are helpful, sometimes they raise legitimate concerns and sometimes they have done harm—for example sparking race riots or pogroms. They can catch fire and spark major changes (for example the abolition of slavery, promoting civil rights), while other iconoclastic ideas remain ridiculed or else mobilise large numbers of supporters but struggle to effect the changes they want to see (for example anti-war or anti-nuclear movements). Social media offer the latest communication tools through which critical ideas emerge, are discussed, developed and disseminated. They open up the discussion of ideas to more people, democratise the debate about policies and political change, enabling users to be the *creators* and *distributors* of their own messages and move them along more quickly. And they can do this for free. They are an inevitable and welcome facet of any open society.

What is good climate science—and who produces it?

Concerns about the extent that social media make constructive climate policymaking harder misunderstand the relationship between science and society. There is a step between how science is constructed and by whom and how scientific knowledge is subsequently accepted or rejected. Science is not a series of 'facts' about the world 'out there' that scientists—and only scientists, preferably clad in white and working in a nice clean lab—discover that are 'true' and that everyone else uncritically accepts are true.

Scientific knowledge is produced, verified and accepted by society—or not—by a series of steps including, but not limited to, the practices of scientists and their labs. What experiments do they decide to do or not do? What do they regard as data to be collected, what is ignored? What methods and experimental techniques do they decide to use or not use? What are their ethical commitments and procedures and how robust and intensive or extensive are their methods? How do they interpret their data and write up their science? Where do they choose to publish their findings (on line, in blogs, peer review journals, open access or not)? How do peer review processes work? These questions, and the answers to them, show that scientific knowledge is socially produced, not unproblematically 'discovered' (Latour, 2004).

Climate science can too often go on inside an opaque 'black box' which non-scientists are not allowed inside. Climate scientists can be proprietary about their work and unwilling to share what they regard as 'their' data. Critics—and increasingly research funding bodies—argue that the processes whereby data was produced and analysed using taxpayer's money, and the data themselves, should be publicly available. Scientists who do not share the knowledge that society has funded can rightly be challenged by critics through social media, uncomfortable though this can be for them. This can lead to better understood and more widely used knowledge about the climate.

There is a series of social processes by which scientific knowledge about climate change is produced, read, accepted or rejected, gains traction in the media or is ignored and challenges or supports common sense ideas of what is or is not appropriate policy. These processes are mediated by cultural mores and by the interests of powerful people and groups. The production of scientific knowledge cannot be separated from the world that interprets how it is received. Through facilitating a wider democratic debate, social media can open up this process of understanding what is or is not robust science and who produces it and why. Social media facilitate debates about what is or is not 'bad' science (e.g. science based on poor data, inadequately analysed, poorly communicated) and enable claims to be aired about illegitimate and ideological uses of that knowledge.

This is a public good. But it is a good that needs to be tempered by concerns about (especially) mainstream media outlets giving equivalence to, on the one hand, the outputs of partisan climate-sceptic groups who disseminate questionable facts on social media with, on the other, more-or-less accepted peer-reviewed science as represented by the IPCC. It was good that radical critics challenged the then-accepted wisdom that the sun goes round the Earth. But there is a difference between being open to radical critiques of science, recognising its normative limitations, and going as far as thinking that 'anything goes'. Some religious fundamentalists still believe Earth was made by God in six days. Others claim the world is flat. Science would dispute that. In this instance we can probably work on the basis that the science is right. Feel free to believe the world is flat, but don't charter a transatlantic flight based on that belief.

Perhaps a middle way is to recognise that scientists can be found in places other than research labs. Citizen scientists can produce their own data using smart phones, photography, personal diaries and logs and produce their own interpretations of what publicly available data means. They can share these data and interpretations though Wiki and open source commons software. These citizen scientists can be a useful part of the process of interpreting what is and is not 'good' science, creating their own robust knowledge independently of or in collaboration with lab-based science. Social media are of immense benefit here.

What is to be done about climate change?

While scientific evidence is clear that the planet is warming, science alone does not tell us what we should do about it (Hulme, 2009). It does not tell us if we should be optimistic or pessimistic about our capacity to make change, how quickly we should do it or whether we need radical, paradigm shifting or evolutionary change. Social media bring more voices to this debate. This is a good thing.

Our understanding of what is or is not a problem, of what should or should not be accepted as an inconvenient but unchangeable reality, is influenced by culture. What does 'too hot' mean between pleasantly sunny on the one hand and people dying on the other? What specific changes or investment decisions should we make or put off, given that we cannot accurately know the future? There are genuine trade-offs between meeting immediate needs and forgoing consumption in anticipation of long-term rewards. We might think that we should prioritise economic growth, address poverty or defeat malaria and leave the climate to richer generations in the future (see **Chapter 1**). Consequently, while Sir Nicholas Stern convincingly showed in his eponymous 2007 report—*The Stern Review on the Economics of Climate Change*—that the costs of mitigation undertaken now would be cheaper than adaptation in the future, coal-fired power stations are still built and new airports commissioned. This is because people are not fully logical and rational beings who understand 'the science' and do what they think it says they should do. Even if it did—and as I have shown—'the science' tells us only that the planet is warming, not what we should do about it.

This is because people evaluate risks differently (Douglas and Wildavsky, 1982). We can be more sanguine about long-term emerging problems than about immediate risks. A lion running towards us will provoke a more immediate response than claims that there will be a famine in ten years if we don't change our agricultural and consumption practices because we think 'something will turn up'. Behavioural economists have shown we have all sorts of biases and rightly so. We are emotional, not logical and science-driven (Thaler, 2015). Consequently, we do not accept what 'the science tells us' and immediately change our behaviour. This is why some of us eat and drink more than what some say is good for our health, why some of us smoke and why some of us drive a car rather than get on a perfectly affordable and accessible public transport system. Science can tell us how much CO_2 a flight emits, but we struggle to understand what such a number actually means in terms of long-term global processes and our individual responsibility. Should we stop flying whilst coal-fired power stations and airports are still being constructed? Or do we think that is a pointless gesture? Science can tell us how much carbon is emitted in constructing, driving and disposing of a particular car, but not whether or not we should individually drive a car.

We have different views about the scale and immediacy of the problem. Rather than being driven by 'science', or by technical understandings of 'what works', people have different ethical and value systems. For some, the need to avoid dangerous climate change requires significant state-led action (see **Chapter 12**), such as proposed by the Green New Deal[1] or a twenty-first century equivalent of the Manhattan or Apollo Projects for clean energy. For others, the capitalist market is the most innovative system humans have ever developed and technological innovation will save us: we elevate what Thomas Friedman (2008) calls Father Profit ahead of Mother Earth (or **Gaia**). Others are convinced that the modernist hubris that suggests we can solve problems through technology is exactly what got us into this mess in the first place. Some see the future looking much like today, but electric and low carbon as a result of the next **Kondratieff Wave** of innovation.

Others still see low carbon capitalism as offensive. In a call for what they call 'a positive future for human freedom', they call for radical system change as climate changes everything (Klein, 2014). Capitalism won't deliver the goods and they won't go quietly into the night. They see the climate crisis as providing the possibility of transitioning to a more democratic, inclusive and convivial economic ethics for **the Anthropocene** (Gibson-Graham and Roelvink, 2010) that will enable us all—humans and other species—to live with dignity and prosperity. Such people see a low carbon future as better than what they characterise as the growth-addicted, consumption-fuelled treadmill too many of us in the Global North are on (Hopkins, 2008).

And then, finally, others see no future for humanity. They argue that catastrophic climate change means that **eco-collapse** cannot be avoided and that we need to manage this inevitable—perhaps even welcome—decline. Some go as far as arguing that Gaia will be better off when it has shaken off a bad case of human infestation, as in the voluntary human extinction movement.[2] Their critics accuse them of advocating misanthropic and possibly racist catastrophism.

Given that science alone can never resolve these ethical and value-based assertions of what should be done about climate change, I argue that new social media are a necessary and welcome technology that can enable us to work through these debates and enhance democratic policymaking.

The facilitation of radical grassroots voices and innovation niches

This leads to my fourth argument: what does 'constructive' policymaking mean? Main-stream conceptions of what constitutes constructive policymaking are well represented through the Sustainable Development Goals or the consensus-based reports emerging from the IPCC. In contrast, advocates of radical, rather than incremental change, are critical of what they regard as these 'post-political' conceptions of inclusive and sustainable development that they believe are inadequate given the immediacy and scale of the problem. Social movement scholars have long recognised that scientists in labs are not the only people producing reliable knowledge. Environmental and climate activists are also 'knowledge producers', developing their conceptions of what is wrong with the climate, and what should be done about it, backed by arguments (of varying strengths) and by data (of varying reliability), thereby resolving the problem (to a greater or lesser extent).

In moving from a conception that 'something is wrong' to 'doing something about it', activists need to mobilise resources from supporters and allies—money, time, media, people prepared to protest, public spaces into which they can put their arguments. Green-peace is the paradigmatic example of a social movement able to raise significant resources in this way. Social media make this easier than ever and—I say it again as it needs to be said—for free. They have enabled activists to construct (often professionally designed) blogs and web pages to raise awareness, develop proposals and demand change. Cultural symbols circulated through social networks can be effective tools of communication, crystallising arguments and channelling action. Web-based crowdfunding and online payments systems offered through social media platforms have made raising money easier than ever.

More significantly, social media offer an increasingly easy way for meetings and pro-test events to be organised. These might consist of global mobilisations—at times and in places where decisions on climate policy are made, such as political meetings of the UNFCCC—to more localised forms of environmental and climate direct action, such as Earth First! protests against fracking and tar sands or opposition to the Dakota Pipeline in the USA. New social movements such as Extinction Rebellion[3] and the school climate strikes[4] are facilitated through social media with protest events documented by those who attended in blogs, web streams and online videos.

ICTs also facilitate prefigurative grassroots or local experiments in and exemplars of what a society in balance with nature could look like. For example, Transition Initiatives[5] are community-led projects developing localised resilient communities. They experiment with local currencies, local food and renewable energy projects, opportunities for local production using local resources and co-operatives and local community-owned businesses. Social movement scholar Alberto Melucci (1989) argues that it is often in such hidden networks that new ideas are developed and emerge which are later accepted more widely. Seyfang and Smith (2007) argue that what they call 'grassroots innovation niches' can act as places where new ways of living can be piloted and tested, before later diffusing into mainstream society. Social media make it easier than ever before for these prefigurative innovation spaces to be organised, find supporters, raise money and other resources and to get their ideas out there. Some of them could be, and are, transformative.

Conclusion

Social media do *not* make constructive climate policymaking harder. Rather, they bring more voices into the debate and facilitate new forms of policy and politics inspired by climate change. Of course we need to develop the capacity of policymakers and citizens to discern good arguments from bad, and well attested facts from fakes. And we need to discern when someone's ideological position (we all have one) distorts their arguments to such an extent that they are not honestly advanced nor backed with reliable and con-vincing data. They may secretly be funded by partisan interests, they may falsify data or wilfully omit data that does not fit with their worldview. Social media—in wired, free societies at least—can call out such egregious deception.

The problem lies more with the volumes of abuse and trolling, not in opening up the marketplace of ideas. Rather than regard social media as a barrier to making good cli-mate policy, we should accept them as offering new tools for critiquing science, under-standing how science is constructed and for bringing new citizen scientists and new forms of data into the debate about what climate change means. Critical, radical ideas about to change society that might actually turn out to be of value can be facilitated through social media. The ideas circulated through a blog about what needs to be done to avoid dangerous climate change can catch fire, as Swedish schoolgirl Greta Thunberg found out in 2018 when she called for a youth strike to demand action on climate change.

Avoiding dangerous climate change is not a closed, technical process that government, industry and science can roll out in uncontentious ways and we all do what we are told. My position is that climate change requires a paradigm shift, a societal transition about which there is no agreement and for which there should be no culture of deference to elite, scien-tific views. Experts, of course, rightly bemoan a lack of public respect for what they regard as helpful, sensible, constructive expert knowledge. But this fails to recognise that avoiding dangerous climate change is a contested political process, not a consensual technical transition. We need as many voices as we can to engage with this challenge, as many people as possible acting at a range of scales and in different knowledge domains.

We need to develop our ability to engage online with people with different points of view, but with civility, learning to disagree well. We need to be accurate and honest about what we write online and better able to discern when people are being deliberately deceptive or supporting hidden interests. We also need to enable those not connected to social media to engage, especially those in the Global South or in theocratic or closed societies. This may be a utopian aspiration. But we should not criticise new social media for their unwelcome side effects when they hold the means for revitalising democracy and bringing people together around new and radical ideas that will be needed to deal with climate change.

Further reading

North, P. (2011) The politics of climate activism in the UK: a social movement analysis. *Environment and Planning A*. **43**(7): 1581–1598.

This paper uses social movement theory to explore the limits and possibilities of climate activism in the UK. It explores why climate activism emerged when it did and how climate activism takes place in a diverse range of political spaces and scales. Yet it remains unclear whether climate activism has the motive power to move to more sustainable ways of organising human society.

Pearce, W., Niederer, S., Özkula, S.M. and Querubín, N.S. (2019) The social media life of climate change: platforms, publics, and future imaginaries. *WIREs Climate Change*. **10**(2): e569. DOI: 10.1002/wcc.569.

This article provides a critical review of the literature on social media and climate change. It shows how social media collapse the 'six degrees of separation' which have previously characterised many social networks and break down many of the barriers to individuals communicating with each other. The authors show how this is having profound effects across society, opening up new channels for public debates and revolutionising the communication of prominent public issues such as climate change.

Schradie, J. (2019) *The Revolution that Wasn't: How Digital Activism Favours Conservatives*. Cambridge, MA: Harvard University Press.

This study of online political mobilisation shows that money and organisational sophistication influence politics online as much as they do offline and casts doubt on the democratising power of digital activism. Digital activism is proving more effective for large hierarchical political organisations with professional staff than horizontally organised volunteer groups.

Sunstein, C.S. (2009) *Going to Extremes: How Like Minds Unite and Divide*. Oxford: Oxford University Press.

This book offers new insights into why and when people gravitate toward extremism. Sunstein marshals evidence that shows that when like-minded people talk to one another, they tend to become more extreme in their views than they were before; for example, when liberals are brought together to debate climate change they end up more alarmed about climate change.

Williams, H.T.P., McMurray, J.R., Kurz, T. and Lambert, F.H. (2015) Network analysis reveals open forums and echo chambers in social media discussions of climate change. *Global Environmental Change*. **32**: 126–138.

This study examines the content of communications about climate change on Twitter and finds that social networks are characterised by strong attitude-based segregation into polarised 'sceptic' and 'activist' groups. Most users interact only with like-minded others, in communities dominated by a single view. The authors suggest that social media discussions of climate change often occur within polarising 'echo chambers'.

Follow-up questions for use in student classes

1. How important are social media as gatekeepers for the information you get? How far does this differ from other people you know?
2. Reflect upon *your* 'echo chambers' and 'filter bubbles' with regards to climate change and climate change politics. Whom do you follow or have subscribed to on social media? What views do you encounter there—and which not?
3. If you encounter incivil posts, comments or other kinds of communication on social media how do you react? How *should* you react?
4. If avoiding dangerous climate change requires global collective action then to what extent can social media can help orchestrate transnational civic responses?
5. To what extent do social media accelerate sustainability innovations at city, town or local scales and so promote grassroots solutions to climate change?

Notes

1 See: https://neweconomics.org/campaigns/green-new-deal.
2 See: www.vhemt.org.
3 See: https://rebellion.earth/.
4 See: https://ukscn.org/ys4c.
5 See: www.transitionnetwork.org.

References

Allan, S. (2011) Introduction: science journalism in a digital age. *Journalism*. **12**(7): 771–777.
Douglas, M. and Wildavsky, A. (1982) *Risk and Culture: An Essay on the Selection of Technological and Environmental Disasters*. Berkeley, CA: University of California Press.
Edgar, A. (2006) *Habermas: The Key Concepts*. Abingdon: Routledge.
Friedman, T.L. (2008) *Hot, Flat and Crowded*. London: Allen Lane.
Gavin, N. (2009) The web and climate change politics: lessons from Britain? In Boyce, T. and Lewis, J., eds. *Climate Change and the Media*. New York: Peter Lang. pp. 129–142.
Gibson-Graham, J.K. and Roelvink, G. (2010) An economic ethics for the Anthropocene. *Antipode*. **41**(S1): 320–346.
Hine, D.W., Reser, J.P., Morrison, M., Phillips, W.J., Nunn, P. and Cooksey, R. (2014) Audience segmentation and climate change communication: conceptual and methodological considerations. *WIREs Climate Change*. **5**(4): 441–459.
Hopkins, R. (2008) *The Transition Handbook: From Oil Dependency to Local Resilience*. Totnes: Green Books.
Hulme, M. (2009) *Why We Disagree About Climate Change*. Cambridge: Cambridge University Press.
Klein, N. (2014) *This Changes Everything: Capitalism vs. the Climate*. London: Allen Lane.
Latour, B. (2004) *Politics of Nature: How to Bring the Sciences into Democracy*. Harvard, MA: Harvard University Press.
Melucci, A. (1989) *Nomads of the Present*. London: Hutchinson Radius.
Painter, J., Erviti, M.C., Fletcher, R., Howard, C., Kristiansen, S., Leon, B., Ouakrat, A., Russell, A. and Schäfer, M.S. (2016) *Something Old, Something New: Digital Media and the Coverage of Climate Change*. Oxford: Reuters Institute for the Study of Journalism.

Pearce, W., Niederer, S., Özkula, S.M. and Querubín, N.S. (2019) The social media life of climate change: platforms, publics, and future imaginaries. *WIREs Climate Change.* **10**(2): e569. DOI: 10.1002/wcc.569.

Schäfer, M. (2012) Online communication on climate change and climate politics: a literature review. *WIREs Climate Change.* **3**(6): 527–543.

Schäfer, M.S. (2017) How changing media structures are affecting science news coverage. In Jamieson, K.H., Kahan, D.M. and Scheufele, D., eds. *The Oxford Handbook on the Science of Science Communication.* Oxford: Oxford University Press. pp. 51–59.

Seyfang, G. and Smith, A. (2007) Grassroots innovations for sustainable development: towards a new research and policy agenda. *Environmental Politics.* **16**(4): 584–603.

Stern, N. (2007) Stern Review of the Economics of Climate Change. London: HM Treasury/The Cabinet Office.

Thaler, R.H. (2015) *Misbehaving: The Making of Behavioural Economics.* London: Allen Lane.

Index

The index covers the debates themselves but not the follow-up questions or bibliographic information (further reading, references). Relevant material appearing in Figures is indicated by *italic* page references; in Tables by **bold**.